rosser, Morto

iesel, the m
e engine

DIESEL

The Man & The Engine

DIESEL

The Man & The Engine

by Morton Grosser

Atheneum 1978 New York

Library of Congress catalog card number 78-6196
ISBN 0-689-30652-0
Published simultaneously in Canada by
McClelland & Stewart, Ltd.
Printed and bound by the Book Press, Brattleboro, Vermont
Designed by Mary M. Ahern
Photo sections designed by Anita Karl
First Edition

For Janet and Adam,
who kept on helping.

Acknowledgments

I owe many individuals and companies thanks for helping with my research on Rudolf Diesel and the Diesel engine. They are mentioned below, less the few who prefer to remain anonymous. A list of acknowledgments provokes the worry that someone has been left out. If they have, I ask their forgiveness, and assure them that I am very grateful for the assistance that virtually everyone I approached gave generously.

My thanks to Airesearch Industrial Division of the Garrett Corporation; Atlas Copco AB; Professor Max Berchtold of the Eidgenössische Technische Hochschule of Zürich; Brown Boveri Corporation; Messrs. Arno Berger, Hubert A. Brugger, and Franz Krieg of Autohaus Brugger, Inc.; Mr. Jan Bonde Nielsen and Mr. Bendt U. Jensen of Burmeister & Wain A/S; Miss Leslie Carlsen of Carlsen Porsche-Audi, Inc.; Caterpillar Tractor Company; Mr. George S. Kent and Mr. Newton B. Drury of Chevron U.S.A., Inc.; The Cummins Engine Company, Inc.; Daimler-Benz AG, and Messrs. Heinz A. Guenther and Arnold B. Shuman of Mercedes-Benz of North America, Inc.; Mr. Robert Davis of Davis Diesel Development; Mr. Richard Scammell and Mr. John F. Straubel of the Engine and Compressor Division of Delaval Turbine, Inc.; Detroit Diesel Allison Division of General Motors Corporation; the Deutsches Museum of München; Dornier Aircraft Public Relations; Electro-Motive Division of General Motors; Dr. S. William Gouse, Jr., of the United States Energy Research and Development Administration; Mr. Richard W. Feder of Executive Motors, Inc.; Fairbanks Morse Engine Division of Colt Industries; Dr. Richard W. Shackson of Ford Motor Company; Mr. H. Martin Gardner of L. Gardner & Sons, Ltd.; Maschinenfabrik Augsburg-Nürnberg AG; Mr. Wolfgang Nothnagel;

Oldsmobile Division of General Motors Corporation; Adam Opel AG; Perkins Engines Group, Ltd.; Automobiles Peugeot S.A.; Roots Blower Division of Dresser Industries; Mr. Edy Sennhauser and Dr. K. Biznia of Sulzer Brothers, Ltd.; Mr. Harold Severance, retired secretary of Standard Oil of California, Inc.; Terex Division of General Motors Corporation; Volkswagenwerk AG, and Volkswagen of America, Inc.

Preface

There are about one hundred and thirty-five manufacturers of Diesel engines listed in the 1977 edition of the *Diesel & Gas Turbine Progress Worldwide Catalog*. (The total is given as an approximation because of the intricate licensing and subsidiary relationships between many of the companies.) A number of these firms are among the industrial giants of the world, and have fascinating histories of their own going back more than a century. The engine itself has been built by the millions, in thousands of different versions over the past eighty years.

Given this background, it will be easy to see why a short book can only present the high points in the history of the Diesel and its builders. That limitation doesn't excuse the author from the necessity of reviewing as much of the field as possible. Many of the sources for this book are technical papers and entries in nineteenth century German archives. Even with charity, a good deal of the prose involved can only be described as turgid, though accurate. Since this book is meant for lay readers, I have tried to distill the accuracy and leave the residue behind. That decision has cost some distortion of the original tone, and resulted in a fairly high density of numbers. I apologize for the former; the latter is essential in any book that treats engines seriously.

A number of companies have been generous with information on the history of Diesel engine manufacture. It may come as a surprise to people outside of engineering to learn how much of the craft guild ethic still survives. Engine builders are more fiercely chauvinistic than governments, especially with regard to their own prized innovations. Inevitably, there are claims and counterclaims of priority (I am speaking now of matters of pride, not mere patent revenue), and there are also independent and virtually simultaneous inventions of the same device or principle in different places. I have tried to sort out the early advances in Diesel design as clearly as the evidence permits, and to include a balanced selection of more recent developments from a true embarrassment of riches. If there are errors, they are mine, and they are not intentional. I will be grateful for verified corrections, and pleased if some deadwood has been cleared away.

M.G.

A Note on Units

All of the measurements and quantities in this book are given in metric units. In particular, they are given in the International System of Units (abbreviated SI in all languages), which is an extension and simplification of the metric system established in France between 1790 and 1801. During the nineteenth century the metric system became the official standard of measurement in many nations, and by 1930, 60 percent of the world's population was using it. Today all of the 144 members of the United Nations except five—Burma, Liberia, two Yemenese republics, and the United States—use metric units. SI is even more coherent and comprehensive than the original French system, and it has been adopted in most metric countries.

There are several reasons for using it in this book. First, the Diesel engine is one of the most international inventions of modern times. Its inventor was born in Paris to parents of part Swabian and part Slavic heritage. He became a naturalized Bavarian citizen while in college, apprenticed in Switzerland, lived for years in France, and later in the capital of a newly united Germany, and considered himself a citizen of the world. His engines were designed in metric dimensions, and the overwhelming majority of Diesels are still designed that way. The drawings for the few that are not—less than 10 percent of present world production—are mostly being changed to SI measurements, or (in some American Diesel engine factories) double-dimensioned during the current transition period.

The second reason for using SI is because it represents a genuine worldwide standard. Quite a few people know that a gallon means a different quantity in Canada than it does in the United States. Not so many are aware that, although horses are horses, a European horsepower and an American horsepower are not the same: like the gallon, they only have the same name. The liter and the watt, the SI units of fluid measure and power respectively, are exactly the same all over the world.

There is of course the problem of habit. This book is being written in California, where the public schools are converting to the SI system, and where many engineering drawings have their inch dimensions in parentheses rather than the other way around. While there is a simple metric-to-U.S. conversion table at the back of the book, there are also a few quick approximations that may speed understanding for those unused to SI units.

The bases of the metric system are standards of length (the meter), mass (the kilogram), and time (the second). One meter equals 39.37 inches, or about 3¼ feet. A man 2 meters tall is a good candidate for a basketball team, and an engine 10 meters long would fill several rooms of a house. The meter is divisible into 100 centimeters, and engine displacement is usually measured in cubic centimeters, a unit of volume 1 centimeter on a side—about the size of a sugar cube. One thousand cubic centimeters make 1 liter, a volume a little larger than a quart milk carton. (For American car buffs, 1 liter of engine displacement equals exactly 61.0237 cubic inches, so the current Chevrolet 305 cubic inch V8 is a five-liter engine.)

The kilogram is a flattering unit for people who are weight conscious. One kilogram equals 2.2 pounds, so fashion models in metric countries often weigh a svelte 50 kilograms. One hundred kilograms would be a healthy weight for a defensive end in football, and 1000 kilograms is a metric tonne, conveniently close, at 2200 pounds, to a U.S. ton.

An important conversion factor in a book about engines is the one relating horsepower and kilowatts. Equivalents for metric and U.S. horsepower are given in the table, but if you are dealing with a reasonable number of them, it is easy to remember that a horsepower is about ¾ of a kilowatt. A kilowatt, conversely, is a very strong horsepower, about 1⅓ horsepower in fact, so that a 100 kilowatt engine is significantly more powerful than a 100 horsepower engine.

Perhaps the other approximations that should be included here are speed and fuel consumption. Twenty-five kilometers per hour is about 15 miles per hour, so 50 kph $\cong$ 30 mph, and 100 kph $\cong$ 60 mph. A truck that goes 10 kilometers on a liter of fuel is getting nearly 24 miles per U.S. gallon. (In Europe, automobile fuel consumption is usually measured the other way around, as the liters of fuel used per 100 kilometers driven. A car that will travel 100 kilometers on 8 liters delivers a respectable 30 miles per gallon.)

The table on page 142 includes more comprehensive and exact conversion factors. After a little practice, you will probably find SI units as easy to assimilate as the billions of people who use them every day.

Introduction

The trunk lid of an automobile is the last thing you see when it passes, and the part of the car ahead that you stare at in traffic. Because of its location it has become a signature and advertising space for car manufacturers. We are used to seeing company names on the rear of automobiles, and even the names of individual models. Occasionally the same word appears on cars made by different manufacturers to announce some engineering feature: AUTOMATIC, to signal that the car has an automatic transmission, or FUEL INJECTION, which means that fuel is supplied directly to the engine's combustion chambers by a pump instead of through a carburetor. Recently another word has been appearing on cars of different makes, and the odds are that you will be seeing it more and more often: DIESEL.

What does it mean? Three things: First, it is the surname of Rudolf Diesel, a German engineer who was born in 1858 and died in 1913. Second, it has become the name of the engine he invented. Third, it means that the machine or vehicle it appears on is fueled with a petroleum distillate, different from gasoline, called Diesel oil.

Although there has been a sudden increase in the number of Diesel-powered passenger cars built during the past few years, it is obvious from the dates of Rudolf Diesel's life that his engine isn't a new invention. The first self-sustaining Diesel engine ran for one minute on February 17, 1894, in Augsburg, Germany. It weighed hundreds of kilograms and stood nearly three meters tall, but it developed only 9.5 kilowatts.

In the years since then the Diesel motor (as it is called in the country of its origin and most other places) has become more and more widely used. Half of all the trucks in Europe, most American locomotives, and innumerable tractors, earth-movers, air compressors, and construction machines all over the world are Diesel powered. Thousands of ships and boats registered in every maritime nation are propelled by marine Diesel engines. There are more than thirty million kilowatts worth in the United States Navy alone. Diesel powerplants generate electricity for the city of København, Denmark, for hospitals in South American jungles, and for satellite tracking stations in the deserts of Australia. Commercial Diesel engines with cylinders the size of a thimble have been marketed to fly model airplanes, and some stationary Diesels are so large that they have a built-in elevator to carry mechanics six stories up from the bottom of the crankcase to the cylinder head.

Everywhere that people need power, the use and manufacture of Diesel engines is growing rapidly. There are inherent and unique reasons for the engine's success, and most of them lie in the original ideas of Rudolf Diesel.

DIESEL

The Man & The Engine

One

Rudolf Diesel's invention was a kind of engine. The word engine can be traced back to the Latin word *ingenium*, which means natural quality or ability. Before the eighteenth century almost any machine, no matter how simple, and whether moved by muscle power, springs, wind, water, or gravity, was called an engine. The invention of the steam pump—the first one was patented by an Englishman, Thomas Savery, in 1698—began a new era in which the name engine was reserved more and more for the machines we now call *prime movers. The Diesel engine is a prime mover: a machine that transforms the energy stored in fuel into motion and useful work.*

Releasing stored energy is more difficult than using energy that is already available. When the wind is blowing, a toy sailboat will carry your message across the lake, but a lump of coal will only lie black and quiet on the shore. The coal contains plenty of stored energy—more than enough to move the sailboat—but to get its energy out, some energy has to be put in first. This is true whether one burns a few twigs to heat up a teakettle or advances the throttle of a jet engine. In both cases the temperature of the fuel must be raised to enable us to tap its stored energy.

The fuel used in a Diesel engine is oil refined from petroleum. Like other petroleum-based fuels, it has to be heated to release its energy, heated enough to burn or *combust.* Engines that burn their fuels are called, logically, *combustion engines,* and they are divided into two groups depending on whether the fuel is burned inside or outside the engine. The Diesel engine burns its fuel inside, so it is an *internal combustion engine.* As practical prime movers the internal combustion engines are Johnny-come-latelys. For almost two centuries before their heyday, machinery was driven by the other half of the family, the external combustion engines. The most familiar prime mover that burns its fuel externally is the steam engine, and it will help to understand the Diesel if we take a short look at its venerable ancestor.

You have probably seen a steam engine operate. Fuel is burned under a boiler to heat water, the water turns to steam under pressure, and the steam is admitted to the cylinder of an engine to force a tight-fitting piston back and thus produce motion. At first glance it seems simple, and in some ways it is. The steam engine can use almost any fuel that will burn: wood, coal, oil, or natural gas, for example, and it will work over a wide range of speeds and pressures. It was a good engine to begin an industrial age because it could be made fairly crudely and still run. But it also has many disadvantages. Two of the most important ones are its size (while a steam engine itself can be made small and powerful, it needs a large burner and boiler system), and its low efficiency.

Efficiency is a word you are going to hear more about. It was one of Rudolf Diesel's guiding principles, and while engineers have always worried about it, many other people all over the world are going to be concerned with it in the years ahead. Efficiency is a very simple, very important idea: *It is the measure of how much you get out of something for what you put into it.*

For any prime mover, what you put in is energy from fuel; what you get out is useful work. Every machine loses part of the energy put into it, so there is no such thing as a 100 percent efficient engine. (Some of the energy is lost overcoming friction and air resistance, some in the production of unusable heat and noise, to name a few major energy leaks.) An engine that converted half of its fuel energy to useful work would be 50 percent efficient. The steam engine had about two hundred years to develop before internal combustion engines began to threaten its monopoly. During that time it evolved from a simple pump into a complex manycylindered machine that used its steam several times over, and was supplied by elaborate high-pressure boilers. Despite all these improvements, most of the steam engines ever built ran at less than 10 percent efficiency, and even the most sophisticated ones could not exceed 15 percent.

If you have ever watched a large steam locomotive in operation, you will realize what an astonishing statement that is. Locomotives are not the most efficient steam engines, and the apparently irresistible force that starts a train of cars weighing thousands of tons represents less than 1/10 of the energy released from the fuel burned. It makes a lump of coal seem more impressive. And, to people who stopped to think about it, it also seemed unreasonable that for each unit of useful energy converted by a steam plant, nine more went up the chimney or were lost in other ways. By the end of the nineteenth century many engineers, Rudolf Diesel among them, were becoming dissatisfied with the low efficiency of steam power.

Two

Even great engineers are not born thinking about engines. Rudolf Christian Karl Diesel was born on March 18, 1858, in a small Paris apartment at 38 Rue de Notre Dame de Nazareth. His father Theodor was a struggling twenty-eight-year-old leather worker (the Parisians called him a *Maroquinier*) who had emigrated from Augsburg in Bavaria. His mother Elise was the daughter of a moderately prosperous Augsburg merchant; she and Theodor met in Paris in 1855 and were married the same year.

Rudolf was a middle child and an only son. One sister, Louise, was born in 1856, and a younger sister, Emma, a year and a half after Rudolf. When Emma arrived, the Diesels moved to the Rue Gontaine au Roi, where Theodor established a larger leather-working shop in the same two-story apartment that housed his family. He hired several apprentices, and for a brief period employed six workmen. Rudolf's toddler years coincided with this short-lived expansion of his father's business. Theodor was a poor financial manager, but a rigid disciplinarian. His shop worked from dawn to dusk, Monday through Saturday, and it was strictly forbidden for the Diesel children to distract the workmen.

Perhaps because of Theodor's strictness, Rudolf grew into a shy and introverted child. He was not allowed to bring any of his friends home for fear they would disrupt the business; after a while his mother noticed that he had no more friends. Theodor had trouble understanding this, especially since Rudolf was a large, handsome boy. Many adults who knew him as a child described him as beautiful. Still he played more and more alone. He spent much of his time drawing, and it was soon obvious that he had natural artistic talent. He also made the kinds of mistakes that curious boys make: a nearly fatal experiment with the gas distribution system, and a one-way disassembly of the family's prized cuckoo clock.

Theodor's business left him little sense of humor for his inquisitive son. A leather strap was the standard punishment for small infractions, but adventures like the cuckoo clock were dealt with by tying Rudolf to a piece of furniture on Sunday while the family went for its regular outing. Fibs were never compromised; the day after being caught in one, Rudolf arrived at the Protestant school with a large sign fastened around his neck: I AM A LIAR. "But I am also a scholar," he could have said with truth. He led his elementary school classes most of the time, drew superbly, and spoke three languages. (German was the language at home, French at school, and his mother, who had worked as a governess in London, also taught him English.)

Following in the steps of his brilliant older sister, Rudolf found school a refuge from the narrow discipline of the leather shop. He also found another refuge on the Boulevard Sebastopol: the *Conservatoire des Arts et Métiers*. This dingy technical museum was a catchall for obsolete inventions, machinery, and models of every description. Rudolf haunted it during out-of-school hours, and one exhibit in particular fascinated him. It was the oldest self-propelled vehicle in the world, a three-wheeled steam tractor built in 1770 by the French inventor Nicolas Cugnot. By a miracle of neglect this mastodon—Cugnot's second model—had survived a century of revolutions and political upheavals. Rudolf sketched it in his notebook with the care due to something important.

While the Diesel children distinguished themselves at school, their parents struggled. France was also struggling through the 1860s under Emperor Napoleon III. His lavish spending and repeated failures of foreign policy provoked widespread discontent among the French people. The volatile Parisians in particular were quick to express their resentment against any handy target. One convenient target was the growing confederation of states called Germany. The

French hostility was returned with interest, because Prussia was a dominant force in the German confederation, and Prussia had been cruelly overrun by Napoleon I. Many Prussians nursed bitter memories of the French, and as the twenty-five German states jelled into something approaching nationhood, relations between the confederation and France steadily worsened. The Diesels regarded themselves as Bavarians—people with temperaments closer to the French than the Prussians—but to the Parisians they were German, and so were the things they made. By 1869 Theodor was selling the entire output of his shop to the German-owned department store, Kellers of Solingen. He had expanded his line of products to include silk handbags and some toys and games, but he was still having trouble making a profit as 1870 began.

At first the year promised well for Rudolf. He finished elementary school with great distinction, and was awarded a medal for superior scholarship from the *Société pour l'Instruction Élémentaire*. He had hopes that his father would be able to repeat his onetime extravagance and rent a weekend cottage in the countryside of Vincennes, and he was looking forward to beginning *École Primaire Supérieure* in the fall. But on July 19 relations between the French and Germans, aggravated by past resentments and present pride, disintegrated; France declared war on Prussia.

Theodor Diesel was an internationalist, but he realized that for his vulnerable family the war was truly a disaster. The performance of the French armies did not encourage tolerance for resident aliens in Paris. Napoleon was counting on the support of Austria and Italy, the Prussians on the aid of the newly allied South German States. To the surprise and humiliation of France, the three German armies that invaded from the east suffered only one setback, at Saarbrücken, before sweeping the French forces before them in a nearly continuous retreat. By August 28 Paris was crowded with displaced families from the provinces, and the War Ministry announced that all foreigners must be ready to leave the city in three days. Elise packed; Theodor closed his studio and tried to find a loan to pay for their transport. On September 1st the French army under the command of Marshal MacMahon and the Emperor was routed at Sedan. The following day Napoleon III surrendered himself and 83,000 troops to the Germans. The news reached Paris on the 4th, and on the 5th the Diesel family traveled by refugee trains to Rouen and Dieppe. At daybreak the next morning they boarded an overloaded channel steamer, and seven seasick hours later the entire family, fed on some bread and one shared bar of chocolate, found themselves safe on English soil, in a single room of a Newhaven hotel. It was not an auspicious beginning for middle school.

Three

When the exiled Diesels arrived in London, 1870 had almost four months to run. It was already a year that Rudolf would never forget, and it still held a few surprises. The only affordable lodging the family could find was a bare two-room flat in Hoxton. The parents shared the bed, the three children a couch, and the boxes they had used as luggage were the furniture. Theodor tried to find work, but his foreign accent and handcraft skills commanded small respect in an England dedicated to empire and industrialization. A retail store agreed to hire him at a wage of one pound a week. Then the clever Louise became, at fourteen, a teacher of language and music in a private school, and the family could at least eat.

And Rudolf? Rudolf was enrolled in a London school, resumed his studies, and discovered new haunts that made the *Conservatoire des Arts et Métiers* pale into shabby obscurity: The British Museum and the South Kensington Museum (now the Science Museum). He was fascinated by the science and engineering exhibits at South Kensington, and also by the ship traffic on the Thames. The Diesels scrimped, the war dragged France deeper and deeper into disgrace, and Theodor and Elise wished that they had stayed in Augsburg. Their letters home were answered pessimistically; business in Bavaria was very poor because of the war. But one letter brought an offer that was to change Rudolf's life, and, as it happened, the future of the industrialized world.

Theodor's cousin Betty was married to Christoph Barnickel, a mathematics professor at Augsburg's Royal District Trade School. The Barnickels proposed to Theodor that they lighten his financial burden by taking in Rudolf and sending him to school in Augsburg, at least until the war ended and the Diesels could reestablish their business. Theodor and Elise accepted the offer with alacrity. Two months after his first dismal channel crossing, Rudolf had a second one, from Harwich to Rotterdam. This time the sign around his neck gave the names and addresses of his relatives in Germany. Eight days and seven trains later he arrived in Augsburg, exhausted and nursing a severe cold. The Barnickels soon brought him back to health, and Christoph wrote to the Diesels that he and his wife had lost their hearts to Rudolf; he was handsome, modest, intelligent, and the most adult twelve-year-old boy they had ever met.

In a few days Rudolf was enrolled in the three-year curriculum of the *Königlichen Kreis-Gewerbsschule*. By a happy coincidence this technical junior high school provided outlets for all his deepest interests under a single roof. The classrooms were on the second floor of a onetime cloister, and they were augmented by a chemistry laboratory and a workshop equipped with a forge and machine tools. On the first floor was a municipal art gallery, which displayed paintings by Albrecht Dürer, Hans Holbein the Elder, Tintoretto, and da Vinci. Scholastically it was a near-perfect environment, and Rudolf, entering at the bottom of his class, was tenth in it by the end of his first semester.

His life was still far from secure. The Barnickels were warmhearted, but they were not Rudolf's parents and they were not rich. As a visitor without German citizenship he was ineligible for scholarship aid. After France surrendered to Prussia on January 28, 1871, the Diesels had to save every cent against their return to Paris. In July they moved back and began the old struggle again. Rudolf sent them homesick letters and gave mathematics lessons to other students to earn money. By his fourteenth birthday he had decided to become an engineer. He wrote to Theodor and Elise, telling them his plan and vowing to rise to the top of his class. Together with his ambition he had acquired a horror of poverty that was to last the rest of his life.

The normal course of schooling for an engineer in Germany was three years of industrial high school followed by three years at a *Polytechnikum* or engineering college. Rudolf worked determinedly toward this goal. He kept his promise, graduating first in his class from the Royal Trade School, but his parents wanted him to leave school and start earning money. He pointed out to them that even if he apprenticed himself immediately, he would work for three years without earning a cent, and would end with few prospects for the future. They could only see the short view. He spent the next five months in their tenement flat in Paris trying to convince them to change their minds, and during that time his beloved sister Louise died of heart failure. The Barnickels renewed their invitation to have Rudolf live with them and continue his schooling, and the grief-stricken Diesels accepted it.

In the autumn of 1873 Rudolf enrolled in the mechanical engineering program of Augsburg's new industral high school. The Barnickels now lived in a larger house, and Rudolf had his own room overlooking the garden. As before, he thrived on his academic work. One piece of demonstration equipment in the physics laboratory made a deep impression on him: a modern version of the ancient Malayan fire piston. In its aboriginal form this was a closed-bottomed tube of bamboo matched to a tight-fitting wooden rod with a small recess in its lower end. A piece of tinder was placed in the recess, the rod was inserted in the bamboo cylinder and struck sharply to the bottom. The sudden compression heated the air in the barrel and ignited the tinder, which was quickly withdrawn and used for starting a fire. The principle of the fire piston was either brought to Europe from Southeast Asia or reinvented about 1807, when a patent was issued for it in England. During the next few years steel and brass ones were sold in small numbers as fire-lighting devices throughout Europe.

The version that fascinated Diesel was more precise and more dramatic than its progenitor. It looked like an elegant bicycle pump made of thick-walled glass. The cylinder had a removable lower end cap to hold the tinder, and when the piston was rammed home you could feel the tube heat up and actually see the tinder ignite. There is good reason to believe that the most important idea in Rudolf Diesel's life was kindled at the bottom of that glass tube. To his teachers' delight, he was receptive to most new ideas, and in three years he became the youngest student to pass the high school final examinations, with the highest grades ever achieved. The Bavarian Director of Education came to Augsburg, interviewed the Barnickels' protégé, and awarded him a thousand-gulden annual scholarship to the München Polytechnikum.

München's cosmopolitan gaiety had several noticeable effects on Rudolf: his interests broadened to include music and he began to make friends. He roomed with Friedrich Lang, the son of a modest nursery owner, and his other close friends were two upperclassmen, Lucien Vogel and Oskar von Miller. Both of them were to affect his future in significant ways. Meanwhile he continued to distinguish himself at school, taught French, rented a piano with his earnings, and took long walks. And he grew. When his parents moved to München in the spring of 1877, they found a handsome man who towered over them at 185 centimeters, and who had received both his German citizenship and a scholar's deferment from the required three years of military service. He moved into their home, saddened to find his mother and father prematurely aged. Theodor was also a little strange. He had turned to spiritualism after his daughter died, and in München he decided that he could cure certain diseases by his personal magnetism and set himself up as a faith healer.

It is more than likely that Theodor Diesel's erratic career had a powerful effect on his son. As a student engineer Rudolf showed an intense dislike of waste and waste motion. He was appalled—almost personally resentful—when the distinguished mechanical engineering professor Carl von Linde told his class that the best contemporary steam engines were only 6 to 10 percent efficient. "Study the possibility of practical development of the isotherm," was Rudolf's order to himself in his notebook. By his senior year he was preoccupied with improving the efficiency of engines. He wanted all of the 7,500 Calories available from a kilogram of coal, not 10 percent: "It follows that one should convert this 7,500 Calories directly into work without intermediaries. But is that in fact possible? This has to be found out!!"

Finding it out would have to wait until after graduation. As July, 1879 and final examinations came closer, Rudolf began to feel sicker and sicker. His illness was diagnosed as typhus, and it respected neither scholarship nor Theodor Diesel's magnetism. Weak and feverish, Rudolf missed graduation and was bedridden through an epidemic summer, but he survived. His treasured cousin Betty Barnickel and many other Bavarians did not. That October Professor von Linde arranged for him to start as an apprentice at the Sulzer Engine Works in Winterthur, Switzerland, and in January of 1880 Diesel came back to München for his deferred final examinations. His scores were the highest in the history of the school. Covered with glory, he returned to Switzerland to learn how to build refrigeration machines and steam engines.

Four

When Rudolf Diesel graduated from college, steam engines were the royalty of prime movers. They had arrived at that position through the slow, successive development of a few basic ideas. The man who first combined those ideas into a practical working machine was Thomas Newcomen.

Newcomen was born in 1663 at Dartmouth, England, and he has justly been called the father of the reciprocating steam engine. His goal was to pump the water out of mines by steam power; he succeeded because his designs were practical, strong, and simple. A vertical steam cylinder stood at one end of a Newcomen engine, and a pump cylinder at the other. The steam cylinder was built over a boiler and furnace, and the pump cylinder was connected to pipes that ran down the mine shaft. A piston worked in each cylinder, and the distance between the pistons was spanned by an overhead wooden beam pivoted at its center. Each piston was connected to its end of the beam by an iron chain running on a curved hammerhead. The whole machine looked like a large steam-driven seesaw.

Newcomen's engines worked on an atmospheric cycle. At the beginning of the cycle, steam was admitted to the power cylinder, pushing the piston up and lowering the pump piston into its water-filled reservoir. Then cold water was sprayed into the steam cylinder; the steam condensed, a partial vacuum was formed under the piston, and atmospheric pressure drove it to the bottom of its stroke, raising the pump and the water trapped above it.

In considering Newcomen's achievements, it is important to remember the technology that he had to work with. Early eighteenth century boilers were riveted up from small iron plates; cylinders were cast in pig iron and drilled—seldom straight or truly cylindrical—on cannon-boring machines.

Engine frames were built of heavy timbers, and their fittings were hand-forged by blacksmiths. It was impossible for Newcomen to get a steam-tight fit between the pistons and cylinders of his engines; he was forced to cap each piston with a leather flap and seal the cylinder above it with water. Steam pipe connections were packed with rope, and a leakproof joint was a rarity. Despite these problems Newcomen's slow, inefficient engines could pump out Europe's deepening mines better than men, horses, wind, or water power. He lived to see them operating in many countries beside his own.

Modified atmospheric engines with wooden frames were still being used in England a century after Newcomen died in 1729. One example still preserved at Dartmouth produced 4.5 kilowatts, and its thermal efficiency could not have been higher than half of one percent. By the time that engine was modernized in 1821, the great Scottish engineer James Watt had changed the steam engine from an atmospheric pressure machine to a prime mover whose working stroke was based on the expansion of steam. In the process he raised the efficiency of stationary steam engines to more than 4 percent.

Only fifty years later (Rudolf Diesel was thirteen, at school in Augsburg) the most advanced Cornish engines could pump 40 million liters per day, and were housed in gigantic power stations. A typical engine of this type had a cylinder bore (internal diameter) of 2.5 meters, and a piston that moved through a 3.5 meter stroke. The working parts of these huge machines were finished like fine jewelry, and their fluted columns were subjects for architectural criticism. (In fact if you blanked out the subject, it would be hard to tell that some of the nineteenth century essays on steam engines were not about classical temples or Gothic cathedrals.)

Hundreds of mills and factories were powered by less ornate but equally productive steam plants, in a bewildering variety of types and sizes. There were vertical and horizontal engines, grasshoppers and side-levers, simples and compounds.

They drove spinning frames, rolling mills, and whole buildings full of machine tools, and they raised miners and minerals from deep underground, in one representative case fifty-eight times an hour, half a million trips a year.

At first the application of steam power to industry was regarded as a simple evolutionary change. That was demonstrated in many old mills, where steam engines were coupled to existing waterwheels to help them along. But as Watt and other engineers perceived, combustion machines were inherently different from their predecessors. The new engines were independent of wind and weather, high or slack water. Given fuel, they ran and ran; their dependability was monotonous compared with earlier sources of power, and breakdowns became annoyances rather than natural phenomena. The change was correctly called an industrial revolution.

The application of steam power to transport caused an even greater revolution. Although Richard Trevithick built the first steam engine to propel itself and a load of wagons on a metal track in 1804, the Rainhill trials of 1829 are usually regarded as the beginning of practical steam-powered transportation. The one locomotive that stayed the course at Rainhill was the *Rocket*, built by George Stephenson. The *Rocket* was 4 meters long, weighed 4.5 tonnes, and generated steam at 3.5 kilograms per square centimeter. It won £500 for Stephenson from the directors of the fledgling Liverpool & Manchester Railway by making ten trips in each direction over a 2.4 kilometer length of track while hauling a 13 tonne train. Its average speed was about 24 kilometers per hour.

Fifty years later Rudolf Diesel could travel over most of Europe on railways. Trains of 1,000 tonnes were commonplace, and locomotives had been driven at better than 120 kilometers per hour. Thousands of people were suddenly able to travel from city to city, even to live in one place and work in another. Never in recorded history had mobility changed so much in one lifetime. There were probably elderly people riding on trains in 1880 who if they had seen a machine moving under its own power in 1829 would have thought it was the Devil.

Another effect of steam-powered railways was to make land transport of goods cost-competitive with shipment by water for the first time. Marine designers and shippers were just as eager to profit from the new technology, and the fitting of steam engines in ships kept approximate pace with the progress on land. The benefits were regularity of service and higher speeds; the chief drawback was the necessity that a ship carry its own fuel. (In the early nineteenth century it was thought that steamships, though useful for coastal service, would never replace sailing ships on long ocean runs because coal for the engines displaced too much cargo.)

A marine analog to Stephenson's *Rocket* might be the *Curaçao*, built at Dover, England, in 1825 and bought by the Netherlands Navy in 1826 as their first steamship. *Curaçao* was 40 meters long, wooden hulled, and displaced 445 tonnes. She was rigged as a three-masted schooner, and her two Maudslay side-lever engines developed 110 kilowatts each. A pair of paddle wheels drove her at a maximum speed of 14.8 kilometers per hour.

If Diesel had taken an ocean voyage in 1880, instead of a train to Switzerland, he could have traveled on the brand new *Arizona*, the first ship to be called an "Atlantic Greyhound." *Arizona*'s 137-meter iron hull was registered at 5,247 tonnes. Her compound engines were built by John Elder & Company of Glasgow, Scotland, and developed 4,679 kilowatts. Seven fire-tube boilers supplied the engines with steam at thirty times *Curaçao*'s boiler pressure. On her trials the *Arizona* gratified her designers by reaching 32 km/hr, enough speed for the ship to take the mythical Blue Riband of the Atlantic from the White Star liner *Britannic* on her maiden voyage.

All that in only fifty years! No one could ignore the progress that had been made, or the social and economic changes wrought by steam power. Perhaps it was only the engineers and inventors who were troubled by its imperfections. It was far from foolproof; steam pipe failures and boiler explosions occurred with disturbing frequency. And even in 1880, it was still fantastically wasteful. Rudolf Diesel wasn't the first person to chafe at these shortcomings.

Five

The logical answer was to improve the steam plant, and many people tried to do it. Thomas Newcomen would have hardly recognized his rude brainchild in the polished, nearly silent steam engines of Diesel's time. But despite their appearance of sophistication and the hard-won improvements of Watt and his successors, they were still very inefficient. Nineteenth century locomotive designers became so depressed by the thought that $\frac{9}{10}$ of the fuel put through their engines was burned to no purpose that they abandoned the concept of efficiency altogether. Instead they classified locomotives by their fuel consumption: 1 kilogram of coal per horsepower-hour. Such an engine could be called more economical than one which burned 2 kilograms of coal for the same output, without having to recognize that both of them were extremely wasteful.

A few engineers concluded very soon that improvements in steam engine efficiency would probably yield only a small difference between two small numbers. Conveyance was an inherent part of the problem: steam must be made in a boiler and brought to the cylinder to do its work. That meant lost heat, and pipes with bends and connections, each one a potential weak spot. And a boiler is rarely the simple object it seems to be. It has flues, joints, rivets or welds, stays and tubes; more places for trouble to start. The idea of using the cylinder, which had to be made strong and leakproof in the first place, as the site of power generation occurred to many inventors. What is ironic is that the earliest demonstration of this *preceded* the invention of the steam engine. The first known heat engine of any kind using a piston working in a cylinder was invented in 1673 by a Dutch scientific genius, Christian Huygens, and it was an internal combustion engine.

The source of Huygens' inspiration is obvious from a

drawing of his engine. It looks like a muzzle-loading cannon standing vertically, mouth up, and it worked in much the same way. Instead of a cannonball, a sturdy piston slides in the bore, and a cord from the top of the piston runs over a pulley above the engine to a weight hanging outside the barrel. Near the muzzle, on either side of the barrel, are a pair of exhaust valves. To operate the engine, a small charge of gunpowder was placed in the bottom of the cylinder, the piston lowered on top of it, and the powder ignited through a touchhole. The resulting explosion heated the air in the barrel, expanding it and forcing the piston up. (Huygens calculated from his experiments that a ½ kilogram charge of powder could lift a 1,360-kilogram piston 9 meters.) When the piston reached the top of its stroke, it uncovered the exhaust ports, allowing the spent gases to escape through the valves to the atmosphere. The cylinder then cooled down, a partial vacuum was produced in its interior, and atmospheric pressure drove the piston to the bottom of the cylinder again, raising the external weight and doing useful work.

Although Huygens' machine must have been spectacular in operation, it had some practical shortcomings. It ran on only a one-shot, two-stroke cycle, and there were no provisions for rapidly reloading the gunpowder or bringing fresh air into the cylinder for the next explosion. Still, it contained the germ of the internal combustion engine: *A combustible fuel, burned with air in a closed cylinder, driving a piston by explosive expansion.* No boiler, no pipes, no conveyance, and small heat loss; it looked like a good beginning.

Unfortunately it was also an abortive one, and for a very simple reason. Engines are dependent on their fuel, as anyone used to engines and short of fuel soon learns to his discomfort. The form of fuel least difficult to introduce into a cylinder and most efficient to burn proved to be a gas or vapor. Until about 1800 there was no easily obtainable gas or vapor to burn in an internal combustion engine. Steam, on the other hand, was a vapor. Because of this vital difference the development of internal combustion machines toddled along beside steam engine progress like a small boy trying to keep up with a brawny older brother. The I-C engine didn't really threaten to grow up until the third quarter of the nineteenth century. Though overshadowed by the great public strides of steam power, it passed some important milestones along the way.

The single most important one was the availability of a suitable fuel. The new fuel was coal gas, produced by heating coal in an airtight oven. Different coals and heating schedules produced different compositions; a typical coal gas contained 46 percent hydrogen, 40 percent methane, 6 percent carbon monoxide, and small amounts of oxygen, nitrogen, and ethylene. The residue in the oven after the gas was extracted was coke, a high quality, nearly smokeless solid fuel. Coal gas burned in air with a clear bright yellow flame, and that, rather than its heating properties, assured its success in a world lit by guttering and expensive tallow candles.

There had been a few gas lighting experiments in the 1700s, but the real debut of coal gas was engineered with show business flair in 1801, when Philippe Lebon, a French chemist, created a sensation by illuminating a Paris hotel with his gas Thermolamps. James Watt's son Gregory saw this promotional stunt, and five years later Boulton & Watt sold the world's first commercial gas lighting plant. Twenty years later London had 500 kilometers of gas mains supplying more than 50,000 burners, and sixty years later little Rudolf Diesel opened the gas jets in his parents' apartment and nearly blew up the whole family. By that time gaslight, and coal gas with it, was a commercial commonplace.

Lebon was more than an impressario. In the same year as his lighting demonstration, he patented an internal combustion engine fueled with a mixture of coal gas and air. It incorporated so many "modern" ideas that it almost leaves one wondering what was left to foresee. His design specified gas and air mixing in a controlled ratio, a fuel mixture pumped into a closed combustion chamber under pressure, mechanically-actuated intake and exhaust valves, and (perhaps most startling of all, considering the date) electric spark ignition. As far as is known, Lebon never manufactured his engine, but his patent was the first of many for gas-fueled engines taken out in England, France, Germany, Italy, Switzerland, and the United States over the next fifty years.

Only a small number of these designs were actually built, and an even smaller number were sold commercially. The first internal combustion engine to operate continuously was built in 1820 by William Cecil, a 28-year-old Fellow of Magdalene College, Cambridge. Cecil's engine was fueled with a mixture of hydrogen and air in a ratio of 1:3, and ran at 60 revolutions per minute. Although the inventor demonstrated his machine before the Cambridge Philosophical Society, he was apparently more sobered than elated by its success. The same year it was finished, he was ordained as an Anglican priest, and devoted the rest of his life to the Church of England.

Like Cecil, many of the pioneer engine builders were intuitive designers who solved problems by repeated experiment. When Rudolf Diesel's turn came he had to hoe the same laborious row, but he had a priceless advantage: a sound thermodynamic theory. That theory was based on the work of a man who died of cholera at the age of thirty-six, and who published only one slim volume in his life: Sadi Carnot.

Carnot was born in 1796, the son of one of the five Directors of France, and the inheritor of his father's talents as an engineer and mathematician. After an adventurous career as an artillery officer, he became a dedicated student of science, and in 1824 he published his *Reflexions sur la puissance motrice de feu.* This short essay summarized the requirements for the efficient design of all heat engines. Carnot's insight was uncanny; he saw straight to the heart of the generalized heat engine cycle, included everything essential, and omitted everything superfluous. His ideas were expressed in clear, jargon-free writing, and his judgments have been confirmed again and again.

Reflections on the Motive Power of Fire contained, among other things, a statement of what is now known as the First Law of Thermodynamics: *Heat and mechanical energy are convertible to each other, but are never created or destroyed, only changed in form.* Carnot also pointed out that the maximum efficiency of a heat engine—any heat engine—depends only on the temperature difference in the cylinder between the beginning and end of the power stroke. This led him to propose three simple-sounding conditions for an efficient heat engine:

> FIRST, the temperature in the cylinder at the beginning of the power stroke should be as high as possible.
>
> SECOND, the gas or fluid in the cylinder should be cooled to the lowest possible temperature at the end of the stroke.
>
> THIRD, the cooling should occur spontaneously as the gas increases in volume, so that no heat is lost to the cylinder walls.

These ideal conditions are virtually impossible to combine in a real engine. They are listed here because many designers aspired to them, but no one more than Rudolf Diesel. By the end of the nineteenth century Carnot's specifications would be called "the original Diesel cycle."

Six

Its inventor-to-be hadn't heard of the Diesel cycle when he returned to Winterthur in January of 1880. The journey from Bavaria to northeast Switzerland carried with it a change of status: from distinguished student engineer to *blaue Monteur*, the humblest of apprentices, clad in the same blue twill coveralls as hundreds of other shop floor employees of Sulzer Maschinenfabrik. Before his examinations Rudolf's first job at Sulzer had been hand-filing a large screw key, a typical assignment for an apprentice in German-speaking countries. He quickly showed that he was not merely a clean-hands engineer and was moved up to a lathe operator's position. He progressed steadily through the drilling and thread-cutting shops, earning the respect and affection of many of his skilled coworkers. The rapport with machinists and firsthand experience with manufacturing processes that Diesel acquired at Sulzer were to prove invaluable later.

Even allowing for his outstanding performance as an apprentice, Rudolf advanced extremely rapidly. He was clearly under the eye of Professor von Linde and the directors of Sulzer, who licensed many of von Linde's inventions. In a short time he was learning the assembly procedures for von Linde's refrigeration machines, and in March of 1880 he was told that he was being assigned—actually loaned—to the Baron Moritz von Hirsch to supervise the establishment of a refrigeration manufacturing plant in Paris. The arrangement was curious: Professor von Linde licensed von Hirsch to build the factory and sell the machines, Sulzer assigned Diesel to the project, and von Hirsch paid his salary. It was a change from blue collar to white collar, but unfortunately von Hirsch, though fabulously wealthy and a great philanthropist, viewed it as a white collar job with blue collar pay. Rudolf's starting salary was 1,200 francs a year (roughly equivalent to that many current dollars), which meant a return to student dinners.

A year later things looked much brighter. The factory on the Quai de Grenelle was up and running, the Linde compressors were beginning to make the same excellent reputation in France that they had elsewhere, and Rudolf's salary was doubled. He was officially the factory manager, but he also supervised many refrigeration plant installations and acted as a consultant and field engineer as well. It was more than devotion to duty; the ammonia refrigeration cycle had begun to interest him theoretically. Von Linde's compressors were, after all, reversed heat engines, and Diesel was unable to let any heat engine lie unimproved. He decided to build an engine using ammonia vapor as the working fluid instead of steam. He could, he reasoned, superheat the ammonia and obtain a greater temperature difference during the expansion stroke than with steam. According to Carnot, this would result in greater efficiency.

By September of 1881 Diesel's talent and industry had brought him three rewards of success in Paris: (1) Another doubling of his salary; (2) His first patent, for the production of table ice in glass containers; and (3) A mistress, in the form of an American divorcée who had met him in München. Presumably the lady was part of the reason why even with his increased salary Rudolf wrote to his sister Emma in October 1881 asking for a 200 Mark loan to secure his patents. She complied, notwithstanding an accurate intuition about his love affair. Diesel's second French patent, for the production of crystal-clear ice free of air bubbles, was granted a month after the first. When his inamorata left Paris, Rudolf was seized with a passion to visit the United States, but that desire wasn't to be fulfilled until twenty-three years later.

In the meantime Baron von Hirsch, whose investments ranged all over the globe, lost interest in the refrigeration business. He sold the factory to another French industrialist and returned the sales license to Professor von Linde. Von Linde promptly assigned the sales license to Rudolf Diesel, first for all of France, and then for Belgium as well. Diesel now became even busier, managing the factory, supervising sales and installations, developing his ice machine, and starting to build the ammonia engine. He was anxious to find a manufacturer for his ice-making machines and approached the obvious person, Professor von Linde. To his surprise, von Linde was not at all interested. The directors of the Sulzer Machine Works weren't eager to build the machines either. Thinking further back, Rudolf remembered that the large and flourishing Augsburg Machine Works, not far from where he had gone to school, had built the first Linde refrigeration plant in 1873. He wrote to the manager of the company about his ice-making machine, and began the most important friendship of his life.

Aktiengesellschaft Maschinenfabrik Augsburg was already a public corporation capitalized in millions of guilders when Rudolf Diesel contacted them. The company had been founded in 1840 by a prominent Augsburg merchant named Sander, but it became a success under the joint managership of two engineers who eventually bought the factory from Sander: Carl Buz and Carl Augustus Reichenbach. Both men were outstanding in their fields; Buz had been chief engineer of the München-Augsburg Railway, and Reichenbach, nephew of the inventor of the flatbed letterpress, was a leading designer of printing machinery. In 1864 they gave the management of the plant over to Carl Buz's son Heinrich, who combined their qualities of shrewd engineering judgment with an open mind and a warm personality. Under Heinrich Buz's direction the company began a great period of innovation and expansion.

It was a record that Rudolf Diesel must have known as an engineering student. In 1873, the same year that his factory built the first Linde compressor, Buz was able to demonstrate the first German rotary printing press; at 10,000 double-sided sheets per hour, it was ten times faster than the fastest flatbed press in existence. Six years later Augsburg had another pair of revelations: the first compound steam engine in Germany and the first rotary illustration press. When Rudolf Diesel contacted Heinrich Buz late in 1881, Buz not only agreed to produce the clear ice machines, but also invited Diesel to Augsburg to experiment with them. One year later *Maschinenfabrik Augsburg* was producing parts for Diesel's first two ice-making plants. Rudolf, on the other hand, was being rapidly thawed by a young lady he had met in Paris named Martha Flasche.

Martha was the talented and pretty German governess of a well-to-do merchant family named Brandes. As a handsome, musical, trilingual bachelor with a rising income, Rudolf fitted perfectly into Ernest Brandes' dinner parties, but it was the governess rather than the guests that attracted him. He was soon a regular visitor to the Brandes home. In the meantime two French breweries had signed contracts for trial installations of his ice machines, one at Châteauroux and another at Châlons. In May and June of 1883 Rudolf, supervising the installations, wrote passionate letters to Martha while coaxing his machines to produce the promised clear ice. Eventually, both trials were surmounted; the ice machines worked, and Rudolf and Martha were married on November 24, 1883, in München.

The Diesels' first year together produced several unexpected developments as well as the hoped-for blessing of a son, named after his father. Rudolf, Sr. had looked forward to the success of his ice-making machines as a source of income. The machines did indeed work, they were patented solely by Rudolf Diesel, and they did produce income. But there was an important restriction in French patent law that Rudolf had either overlooked or never heard of. He was still technically employed as sales director for Linde refrigeration plants, and all the profit earned by his machines was legally due to his employer. Diesel took the logical course with mixed feelings: he resigned, and became an independent representative for the Linde machines.

A second unhappy surprise was the success of the fanatic French general Georges Boulanger at reviving violent anti-German sentiment in Paris. Beginning about the time of the Diesels' marriage and increasing through the next six years, Boulangism affected all classes, resulting in the cooling of many of Rudolf's French friendships and business contacts. Of course he had seen it all before, but now he spoke the language flawlessly and was for the most part accepted as a native. He hung on, trying to sell refrigeration plants and ice-making machines against increasing French resistance to German or even Swiss-made products. His income dwindled, and a symptom which had troubled him as a boy returned in more virulent form. During his family's exile in London he had begun to have violent, long-lasting headaches. Now they started again, worse than ever; he was to suffer from these excruciating migraine attacks at times of stress for the rest of his life. A daughter, Hedy, was born in October of 1885, adding to his happiness and financial worries.

Perhaps it was the promise of the Paris Exposition of 1889 that kept his hopes up. All through the summer of 1888 he worked man-killing days, sometimes twenty hours at a stretch, on the ammonia engine. It was a cantankerous beast, and every tiny leak of the acrid fuel brought nausea and burning eyes and throats to its constructors. Diesel hoped to show it at the Exposition, where he was also licensed to exhibit the Linde refrigeration machines. By the beginning of 1889 it seemed that he had won through: the ammonia engine ran, and it was as predicted more efficient than steam engines of the same capacity. But it was also tricky and dangerous to operate. On May 3, 1889 the Diesels' third child, Eugen, was born. Rudolf decided to exhibit the Linde machines, but not his engine at the Paris Exposition.

Seven

There is good reason to think that 1889 was a crucial year in the formation of Diesel's ideas. On March 31 the *Tour Eiffel* was completed, signalling the opening of the Exposition from the record-breaking height of 300 meters. Every important engine builder and many obscure ones as well seemed to have shared Diesel's ambition to show their inventions at Paris. There were, of course, steam engines: stationary and marine, single, compound, and now triple-expansion engines, larger and more impressive than ever. There were steam locomotives from many countries, boasting new wheel arrangements, new firebox designs, and higher tractive efforts. Steam power was obviously in its prime. But now, not quite so inconpicuous, not always pushed into remote pavilions, there were other prime movers as well.

The availability of coal gas had sparked a surge of experimentation during the years following Carnot's publication of his *Reflexions*. Between 1824 and 1860 a series of English, French, and Italian inventors built gas-fueled engines that worked on variations of the cycle first patented by Philippe Lebon. The first production internal combustion engine in the world was patented in Paris in 1860 by Jean Joseph Etienne Lenoir. About 500 Lenoir engines were built, mostly of 2 kilowatts or less. They burned a 7 to 1 mixture of air and gas, and they were afflicted with an electric ignition system of extreme delicacy.

Diesel was doubtless familiar with the Lenoir engines and the successors to them designed by Pierre Constant Hugon and Alexis de Bisschop (a Parisian despite his un-Gallic name). The Bisschop machines were air cooled, one cylinder atmospheric engines of low power ratings—250 watts and less—but they were ingenious, cheap, and reliable. Their ignition system was a simple pilot flame behind a flap valve uncovered by the piston. A 62 watt Bisschop engine (billed as "one manpower") sold for the equivalent of about $125 in 1878, and burned five cents worth of fuel an hour. The only maintenance required was oiling the bearings, and Bisschop engines were reported to have run for more than a month at a time nonstop. They are unknown today except in technical museums, but they filled a need that grew with them, proliferating in the shadow of contemporary steam plants. By 1889 they had been in continuous production for more than twenty-five years in England, France, and Germany, and had sold by the thousands.

Not much attention would have been paid to Bisschop's designs at the Paris Exposition. The real news for believers in internal combustion came from Germany. The men responsible for it were the founding fathers of the internal combustion age: Nicolaus August Otto, Eugen Langen, Gottlieb Daimler, and Wilhelm Maybach.

Otto was the one who started it. Like Diesel, he was intrigued, or rather obsessed with engines. Again like Diesel he came from a struggling merchant background, and though his widowed mother recognized his technical talent when he was a boy, she urged him to find a career in commerce. There the parallel ended, because Otto had no kind cousin to offer him schooling or a roof over his head. He had to leave school to earn money, and became first a grocery clerk and then a traveling salesman for tea, sugar, and kitchenware. Otto was twenty-eight years old when the Lenoir gas engine was patented, and it was a powerful spur to his suppressed engineering ambitions. His sales territory lay along the German side of the French and Belgian borders but did not include Paris, where the Lenoir engines were made. In 1861 he had a Köln instrument maker build him a small working model of a Lenoir engine, and from that time on every spare moment and all of his funds went into the effort to build a better internal combustion engine.

Two years, three models, and much aggravation later,

Otto completed an atmospheric cycle gas engine that ran more or less continuously. It contained some original ideas, most of them stemming from Otto's preoccupation with piston shock (his second engine had pounded itself to pieces), but it could not have been called a startling advance over the Lenoir, and it had drained almost all of the inventor's savings.

On February 9, 1864, Otto was a participant in one of those storybook encounters so beloved of screenwriters. That afternoon a young Köln engineer and businessman named Eugen Langen heard the irregular pounding of Otto's engine in the back of a small shop on the Gereonswall. Langen was already a success in several ways; from his student days at the Karlsruhe Polytechnic he had gone on to become a partner in his family's sugar refining business, director of a steel forging plant, inventor, and design engineer. At thirty years of age he considered himself underinvolved. He went into the shop to investigate the source of the noise. Otto's engine fascinated him, and the personalities of the two men formed a rare complementary blend. In a few weeks of enthusiastic work Langen raised the capital to form N. A. Otto & Cie on March 31, 1864. It was a cornerstone well-laid; today, after a series of renamings, it is Klöckner-Humboldt-Deutz AG, the oldest manufacturer of internal combustion engines in the world.

Despite Langen's optimism, success didn't come easily. It was a familiar story: the prototype, which seemed to need only a few refinements, actually took three years of grimy, disheartening work to evolve into a practical engine. And although Langen was nominally the business partner, his engineering designs were also crucial to the engine's success. The Otto and Langen atmospheric engine of 1867 was very different from Otto's original machine. It was also one of the most peculiar-looking prime movers ever designed, a factor that weighed against it when it made its debut at the Paris Exposition of 1867. Its single vertical cylinder was cast in the form of a fluted Grecian column, and it stood on a paneled octagonal base concealing the valve gear. The flywheel and transmission gears were exposed above the cylinder, giving the engine the appearance of a large coffee grinder on a classic pedestal.

As if that were not enough, it made a horrendous racket when running. The judges at the exposition, most of whom were French, politely looked the other way and concentrated their attention on the designs of Lenoir and Hugon. But one judge, Franz Reuleaux, had been a student at Karlsruhe with Eugen Langen. He suggested to the board that they make their awards on the basis of efficiency as well as apparent smoothness of operation. In two days of testing the panel found that the Otto and Langen engine used less than half the fuel of the best competing engine of the same output. This remarkable performance won Otto and Langen a gold medal, the grand prize, and more orders for their engines than they could possibly fill.

Back orders meant more manufacturing plant, and that meant more capitalization. In a series of hurried expansions Otto & Cie tried to keep up with its growing reputation. Finally in January of 1872 Langen, his brothers, and several partners reorganized the company as Gasmotoren-Fabrik Deutz AG. (The name came from the Deutz suburb of Köln where the new factory was located.) Otto was technical director, and a few months later the company hired Gottlieb Daimler as production manager; Daimler brought with him a young engineer named Wilhelm Maybach. Daimler drove a hard salary bargain, but any director of personnel who could duplicate this recruitment today would deserve a generous bonus. The fledgling Deutz Gas Engine Manufacturing Company now had under its roof three of the greatest engine designers who have ever lived.

These brilliant engineers acted as mutual catalysts. Maybach redesigned the atmospheric engine, simplifying and improving it. Daimler contributed ideas—not always without friction—and incorporated the changes into improved production methods. Daimler was a technocrat who demanded the highest quality in manufacturing. The engines reflected his perfectionism, and in a short time prospective licensees from every industrial nation were lining up to sign manufacturing contracts for Deutz designs.

Otto, secure for the first time in his life, had time to think about basic principles. Early in his design career he had been interested in the idea of compressing the gas in an engine cylinder before ignition. Inadequate machining facilities, lack of time and money, and the need to produce a well-tried design made him set that project aside. All of his engines, like others on the market, ignited the fuel at atmospheric pressure. But mixture compression appealed to him for several reasons: First, it was a way to reach a wide temperature difference in the cylinder, because compressing the gas raised its temperature. According to Carnot's criteria, this would raise the efficiency of the engine. Second, the presence of a compressed mass of gas in the cylinder could act as a cushion, smoothing out the jarring detonations and abrupt piston reversals in the atmospheric engine.

Piston shock was the bogeyman in Otto's design closet. It is clear from his writings that he "felt" the gas mixture in the cylinder as a palpable volume. One day while watching a factory chimney he noticed that the smoke column, at first opaque and compact, became progressively less dense as it rose higher. The rising smoke, like Newton's falling apple, triggered Otto's imagination. Visualizing the column as the contents of an engine cylinder, he hypothesized that if the gas mixture were rich in fuel near the ignition port and progressively leaner toward the piston, the combustion would spread smoothly through the charge, lessening near the piston. Even the leftover burned gases from the previous detonation could play a part in this stratified charge, forming a "pillow" at the piston face.

Otto tested his hypothesis in a way that will sound

familiar: In 1872 he built a hand-cranked model engine with a transparent glass cylinder, so that he could watch the behavior of smoke drawn into it. His experiments led him to a conclusion that provoked skeptical amusement from many of his contemporaries, and cool disbelief from Daimler. The four stages of intake, compression, ignition/expansion, and exhaust would each be assigned an individual piston stroke, and the whole cycle would require four piston strokes or two crankshaft revolutions. It is easy to disparage Otto's critics, but their skepticism was reasonable. Every engineer was concerned with efficiency. It seemed logical that for high efficiency each revolution of an engine should earn its keep by producing power. The idea of throwing away three strokes out of every four looked on the face of it naive and wasteful.

Eight

Otto was persistent. Though trained at a Polytechnic like Daimler, Langen had seen Otto's intuition triumph before. He assigned Otto an assistant engineer and left him alone, one of the great managerial decisions of the nineteenth century. In May of 1876 the world's first four-cycle internal combustion engine ran—and ran well. Otto's confidence in mixture compression and the four-stroke principle was vindicated immediately, even in the one-cylinder, flame-ignited prototype. The test machine was handed over to Wilhelm Maybach for development into a production model, and the result was the Deutz "A" engine of 1877.

This design was a quantum jump in prime movers. A Model A four-stroke engine weighed 75 percent less than the Otto & Langen atmospheric pressure engine of the same rating, and produced equal power from 1/15 the piston displacement. Indicated thermal efficiency was 16.7 percent, far above contemporary steam plants. The compression ratio of the prototype was only 2.5 to 1, yet its noise and vibration were small fractions of what they had been in the noncompression engines. (The new model was advertised as the "Silent Otto.") It was the engine that began the age of internal combustion. By the time of the Paris Exposition of 1889 Deutz had built 8,300 four-strokes, their English licensee more than 5,000, and their Philadelphia subsidiary claimed sales of more than 25,000.

The tremendous initial success of the Otto four-stroke engine was not entirely due to its efficiency, or to Daimler's concern with manufacturing excellence. The patents that Otto filed for and received assured Deutz AG a virtual worldwide monopoly on all four-cycle engines, as well as on any engines which, in Deutz's judgment, did not extract all the burned gases from the cylinder by the end of the exhaust stroke. The company was assiduous in sueing infringers of the Otto patents, at first with awesome success. But their obsession with legal rights was ultimately a mixed blessing. (In fact that seems to be a recurrent theme in the history of engines; for men used to working with the solidity of iron and steel, paper has often proved to be a fickle medium.)

Internal legal disagreements began early at Deutz, when Daimler demanded, and received, the ownership of patents for ideas originated by Maybach. By 1882, when court battles over the Otto patents were already frequent, Daimler became disaffected. He left Deutz to begin research on a lightweight, higher-speed internal combustion engine. Maybach joined Daimler within a few months, compounding the loss to Deutz. Meanwhile Otto & Langen's relentless grip on the four-cycle engine provoked a whole new field of competition. If mixture compression worked so well in a four-stroke, why not apply it to the two-stroke cycle? The time may not have been right nor the technology adequate for that idea, but its development was forced by the stranglehold of the Otto patents.

In England, Germany, and the United States, engineers took another look at the concept of producing power at every revolution of the crankshaft. A thorough and perceptive Scotsman, Dugald Clerk, laid the foundations of two-cycle compression design with an engine patented in 1878 and demonstrated at Kilburn, England in 1879. Despite opposition from Deutz, Clerk continued to work on two-cycle engines during the next decade. His lucid analysis of internal combustion thermodynamics was to contribute to the development of both spark ignition and Diesel engines for many years.

In Germany life was much more difficult for engine developers in the early 1880s. The threat of Deutz lawsuits (in several of which the loser had been ordered to pay damages, all court costs, and his customers levied with royalties to Otto) discouraged public disclosure of new designs.

Several viable ones were produced even in the face of such pressure. The most important one was the little one-cylinder engine built by Karl Benz in 1879. Benz followed Dugald Clerk's ideas, but he eliminated the auxiliary charging cylinder Clerk had used and many ignition problems with it. In October of 1883 Benz founded *Benz & Cie* in Mannheim, which manufactured the best-designed two-stroke engine of the early 1880s. At that point Deutz's influence was so great that the German courts refused to grant Benz a patent on his engine despite the originality of his designs. (The same engine was granted patents in England and the United States.)

Two-stroke compression engines had some inherent problems compared to four-cycle engines of the same displacement. Everything had to happen more quickly: the time to get unburned gas into the cylinder and exhaust gases out was less than half that in a four-stroke. The ignition frequency was twice as high, a difficult barrier in the days of flame ports and primitive magnetos. Two-strokes ran hotter and used more fuel for a given power output. (Tests on an 1885 Clerk engine showed a brake thermal efficiency of just under 13 percent—about 4 percent less than an Otto four-stroke of the same period.) All of these disadvantages mattered little after January 30, 1886, when Otto's basic German patent was overthrown in a court case publicized around the world.

In 1884 a patent attorney for Gebrüder Körting, gas engine manufacturers in Hannover, found what many other people had been looking for, a prior patent describing the Otto four-stroke cycle clearly and unambiguously. It had been filed in France on January 16, 1862, by Alphonse Beau de Rochas, a transportation engineer for the French government. The Beau de Rochas patent included no drawings, and its author not only failed to build the engine he described, he never paid the patent publication fees or the taxes required to keep his patent in force. Nevertheless, in eight succinct descriptive conditions, Beau de Rochas demolished Otto's claim to prior invention. Here they are, in Dugald Clerk's contemporary translation from the French:

> The most efficient explosion engine must have:
> 1. The greatest possible cylinder volume with the least possible cooling surface;
> 2. The greatest possible rapidity of expansion;
> 3. The greatest possible expansion; and
> 4. The greatest possible pressure at the commencement of the expansion.
>
> This leads to the following series of operations:
> 1. Suction during an entire outstroke of the piston.
> 2. Compression during the following instroke.
> 3. Ignition at the dead point and expansion during the third stroke.
> 4. Forcing out of the burned gases from the cylinder on the fourth and last return stroke.

Otto himself would have had difficulty writing a clearer description of his engine's principles. *Körting v. Otto* ended in a total defeat for Deutz AG. Although Otto's claims were upheld in a British suit, and the American version of the case never came to trial, the dam had burst. Benz & Cie immediately switched to four-cycle engine design, followed within several years by almost every other two-cycle engine manufacturer. Karl Benz, freed from the spectre of Deutz litigation, revealed that his basic drive had been not merely the development of a better internal combustion engine, but the invention of a practical internal-combustion-powered vehicle—the automobile.

Daimler and Maybach had secretly been working on exactly the same project. The independent efforts of Daimler and Benz required the solution of many problems, first among them carburation, the proper mixing of air with the gasoline that they both recognized as a superior fuel for their engines. The second major problem, ignition, took a great jump forward with the magneto designs of Robert Bosch. In July of 1886 Benz's three-wheeled "Dogcart" ran on a public road; in September of the same year Daimler and Maybach's modified horse-drawn carriage ran on the garden paths of Daimler's estate. There was no looking back.

Nine

Rudolf Diesel was aware of these developments in great technical and political detail. The exhibits at the 1889 Paris Exposition were the portents of a revolution in prime movers. Benz & Cie showed the first publicly exhibited automobile, their three-wheeled *Patent-Motorwagen* (Benz himself was hard at work on the more advanced four-wheeled *Viktoria* to be introduced in 1893). Daimler and Maybach, still unincorporated, did not show their motorwagen but their *Standuhr* ("Upright Clock") engine, nicknamed for its shape, ran continuously on exhibition. Its outstanding reliability, together with Daimler's salesmanship, was to make it the powerplant of choice in nearly every pioneer French auto. Daimler and Maybach had set out to build a lightweight internal combustion engine that would run at triple the speed of Otto's original four-stroke design. The *Standuhr* did exactly that, and together with its descendants, began to find applications everywhere. One of them stood on rails outside the engine pavilion at Paris: a tramcar, powered by a Daimler engine.

The engineering news was clearly coming from Germany; and Diesel, with his German and French connections and his solid engineering background, was a unique resource in Paris. When the International Engineering Congress held in conjunction with the 1889 fair was convened in September, Diesel was the only German national invited to address the assembly. His paper, delivered in perfect French, was titled "*Revue Technique de l'Exposition Universelle.*" It is hard to tell how much of its warm reception came from the audience's acceptance of him as a Frenchman. By this time the fanatic Boulanger had been elected to the French national assembly, and anti-German sentiment in Paris was so extreme that several of Diesel's close acquaintances questioned the wisdom of his marriage to a German girl.

But Rudolf was German by descent. Engineering history was being made in Germany, and it was nearly impossible to sell German-made machines in France. The next step was obvious. In November of 1889 Rudolf went to München for a talk with his old mentor Professor von Linde. Von Linde was still a friend, and he agreed with Rudolf's conclusions. When Diesel returned to Paris, the French refrigeration franchise, which had brought him to the edge of bankruptcy, was exchanged for the sales rights for north and east Germany, and a position with a guaranteed annual salary of 30,000 francs. The new job was headquartered in Berlin, the fast-growing capital of unified Germany; the Diesels moved there in February of 1890.

Berlin was very different in tone from Paris. Martha loved it, but Rudolf had a hard time adjusting to a social life dominated by Prussian military families. Still, the city was technical-minded to a fault. The streets were scrubbed clean, locomotives in the green and gold livery of the Prussian State Railways left the six depots for destinations all over Europe on tight schedules, and here and there magical electric lights signalled the presence of the cables recently laid underground by the *Siemens-Werke*. Engineers definitely commanded respect in the Prussian capital, and almost the first thing that Diesel did after renting a new apartment for his family on the fashionable Kurfürstendamm was to set up a workshop for engine research.

The ammonia engine was still on his mind, and still not practical, but working with a dangerous gas at high pressure led him to a crucial conceptual leap. He decided that what really mattered in a heat engine were high temperature and high pressure. If you could raise those quantities to extreme levels before the power stroke, you could produce a very efficient engine with any working fluid—gas or vapor—even ordinary air. Perhaps it was at this point that he remembered the glass fire-piston from his Augsburg schooldays. In any case, during the next year, while establishing the Linde franchise on a sound basis and acting as newly elected director of a refrigeration sales association, Diesel formulated the concept of his *Verbrennungskraftmaschine*—a combustion engine that would surpass all others in efficiency, based on the extreme compression of ordinary air.

A chance encounter with an old friend stimulated him to turn his idea from theory to practice. On a business trip to Frankfurt in 1891 he visited the International Electrical Engineering Exhibition. The president and technical director of the exhibition turned out to be his München Polytechnic schoolmate Oskar von Miller. Miller was a pioneer in the new field of electrical engineering, and when Diesel met him in Frankfurt he was in the midst of a meteoric career as cofounder and director of German Edison Company (later A.E.G., the German General Electric). Miller proudly showed Diesel his latest project, a 16,000 volt electric line connecting a Siemens turbo-generator at the falls of the Neckar River near Lauffen with the Frankfurt exhibition 175 kilometers away. At the Frankfurt end the line was to power a record-breaking 73.5 kilowatt electric motor.

Diesel and Miller had been friendly rivals at school. Miller's electric motor reawakened Diesel's competitive urge. After all, it was only a converter of energy, not a prime mover like his proposed combustion engine. Diesel went home to Berlin and spent the next six months working on the theory and design of his engine. He applied for a patent, but the Imperial patent examiners rejected his first application as "not original." On appeal they changed their minds and issued a developmental patent on what was to become the Diesel engine. The date of the certificate was February 28, 1892.

Like many other internal combustion designs before it, Diesel's concept was a "paper engine." With the promise of patent protection for 15 years, he set out to find a factory that would convert his software to hardware, and followed the same sequence that he had with his ice-making machines.

As before, von Linde, first to be contacted, was not interested. In March 1892 Rudolf wrote to the Augsburg Machine Works; he was optimistic, not only because Heinrich Buz had been so receptive to him earlier, but because Lucian Vogel, his other close school friend, had married Buz's daughter and was working as an engineer at the Augsburg works. To his surprise and disappointment, Buz turned him down. There were two reasons, the first of which was beyond Diesel's control. Josef Krumper, the factory's chief engineer, was a steam engineer and a bastion of conservatism. Krumper could also point to the proven success of his new triple-expansion steam engines when he counseled Buz to reject the proposed heated-air combustion engine, which had only a theoretical basis.

The second reason for Buz's rejection requires a look at what Diesel was proposing to build. Diesel's patent was based on an autograph manuscript finished early in 1892: *The Theory and Construction of a Rational Heat Engine to Replace Steam Engines and Contemporary Combustion Engines.* (Not a title likely to warm a steam engineer's heart.) Diesel started by outlining five possible cycles of combustion including the three most important ideal cycles:

1. Combustion at constant pressure
2. Combustion at constant volume
3. Combustion at constant temperature

Although he was convinced that the key to high efficiency was high pressure, he was also a confirmed disciple of Carnot. He specified the third cycle, combustion at constant temperature, as the theoretical operating principle of his engine. The quest for this unattainable goal was to lead him up several blind alleys.

Still, there were other engineers who had aspired to ideal temperature cycles. What stopped Heinrich Buz was the pressure requirement. Here is how Diesel visualized the four-stroke cycle of his engine:

1. INTAKE: A charge of pure air at atmospheric pressure is drawn into the cylinder.
2. COMPRESSION: The piston compresses the air to 250 atmospheres—about 253 kilograms per square centimeter. During this compression the temperature in the cylinder rises to approximately 900°C, far above the ignition temperature of the fuel.
3. POWER: As the piston begins its outward stroke fuel is injected into the cylinder. It is ignited instantly by the superheated air, and burns at a rate such that the heat of combustion is entirely absorbed by the expanding volume in the cylinder. At the end of the power stroke the pressure in the cylinder should, ideally, be back to atmospheric, and no heat should have been lost to the cylinder walls.
4. EXHAUST: The spent gases are ejected from the cylinder as the piston returns to the intake position.

The cycle sounded feasible in theory, and Diesel calculated that its potential benefits were staggering: If it could be made to run, its theoretical efficiency would be between 70 and 80 percent. He was candid enough to warn that such fantastic figures weren't likely to be achieved with any real engine. He was also aware that half of those values, a reasonable goal, would yield triple the output of any existing steam engine from the same fuel.

In his original manuscript Diesel suggested an upper compression limit of 300 atmospheres. As far as was known in 1892, pressures of that magnitude only occurred in volcanoes and bombs; the highest pressure that had been measured reliably in any man-made device was 37 atmospheres or 38.67 kilograms per square centimeter. In an effort to make his engine look more feasible, Diesel reduced the working pressure to a mere 150 kilograms per square centimeter when he wrote to Heinrich Buz. Buz was an open-minded but practical man; he said no.

Diesel realized that if Buz wouldn't accept these pressure levels no one else would either. He recalculated the essential cylinder conditions at the end of the compression stroke, and came up with an absolute minimum pressure of 44 kilograms per square centimeter. The new figure went off to Buz. Martha Diesel was well aware by this time that she was married to a compulsive worker with an obsession. She sustained Rudolf during the agony of waiting for a reply, and suggested that since the high-compression engine was protected by a patent, why not publish a description of it to attract potential builders? Diesel thought it was an excellent idea and began to rewrite his manuscript into a short book, *Eines rationellen Warmemotors*. On April 20 the good news came: Buz wrote back to say that Maschinenfabrik Augsburg would build an experimental engine to see if Diesel's proposed cycle would work.

Both in his own time and in retrospect, Heinrich Buz stands as a model of enlightened industrial management. He was a solid engineer, not a gambler, yet he undertook the development of many new inventions at early stages of their existence and brought them to successful production. He was also loyal and persistent; once he accepted that a new idea was feasible he backed it steadfastly. His plan for the experimental Diesel engine, worked out during the summer and autumn of 1892, exemplified these qualities. A laboratory-workshop would be set up at the Augsburg works where the engine would be built and tested under Diesel's direction. Each stage would be evaluated before proceeding to the next one, but as Buz demonstrated later, there were to be few limitations of time, money, or encouragement.

Ten

Eight days after he received Heinrich Buz's letter of acceptance, Diesel had his patent attorney submit a second petition to the German patent office. Probably because he was worried about acceptance of his ideas, he deleted two of the broadest claims that had appeared in the first application. These claims were integral to the concept of the Diesel cycle: First, that air and fuel were not mixed before induction into the cylinder; and second, that the air was compressed to a level where its temperature far exceeded the ignition point of the fuel. Although the patent examiner restored the sense of both claims in prefaces to the description of the constant-temperature cycle, the change cost Diesel some aggravation later.

Patent applications were also filed in England and America at about the same time. Diesel worked on his book manuscript and worried that the contract with Augsburg would not be adequate to fund the development of the engine. (Buz was worried about the same thing.) *Eines rationellen Warmemotors* was published in Berlin in January 1893, and received immediate and severe criticism in the German scientific press. Diesel's theories did find a few prestigious defenders; they were either friends, or men with practical experience in building and testing engines, including von Linde, Eugen Langen of Deutz, Professor Moritz Schröter of München, and Gustav Zeuner, the director of the Dresden Polytechnic.

To his credit, Buz was undeterred by the attacks, and on February 21, 1893, Diesel signed the contract with Maschinenfabrik Augsburg. With Buz's help and blessing, he then contacted the Friedrich Krupp Werke at Essen, asking if they wanted to share in the development and eventual profits of his engine. Diesel was a good salesman, but Buz's reputation was even more persuasive. On April 25 Krupp and Augsburg agreed to divide the costs of producing a salable model of the new prime mover equally between them. In exchange for its investment Krupp received exclusive sales rights for stationary, marine, automotive, and locomotive Diesel engines in the German territories not already reserved by Augsburg, all sales rights for Austria-Hungary, and 33 percent of the profits.

The third partner in the agreement was the inventor, Rudolf Diesel, and his share makes eye-opening reading. He was to receive a salary of 30,000 marks annually (which would allow him to devote all his time to the engine's development) during the initial testing period, followed by royalty payments of 37.5 percent of the factory prices of engines sold until he had received 500,000 marks, after which his royalties would be 25 percent of the net sales price. To the ex-struggling refrigeration salesman the contract must have looked like a scroll out of Grimm's fairy tales. It is interesting to consider what one would have to invent today to persuade, say, General Motors and U.S. Steel to sign such a joint contract.

But Diesel was a nonstop worrier. The prospect of a handsome salary for working full time on his own invention, and the promise of great wealth if he succeeded produced even more than usual apprehension. It seemed to him that his family was already capable of absorbing the increase in his income; there was a resurgence of the wretched headaches. He drove himself to supply the Augsburg works with engineering drawings for the test engine, and Buz moved ahead at high speed. Buz showed the value he placed on the project by appointing his son-in-law, and Rudolf's friend, Lucian Vogel as engineering assistant. Vogel was a good choice; he believed in the engine from the beginning.

In May 1893 Sulzers of Winterthur, who had been somewhat reserved when Diesel sent them a copy of his book in January, offered to pay him 20,000 marks a year for an option on the Swiss patent rights. He signed that contract on May 16, and shortly thereafter applied for a supplemental German patent, specifying combustion at constant pressure and other features that were closer to those of the eventual Diesel engine. On July 17 he moved to Augsburg to begin the first series of tests.

When Diesel arrived at the Augsburg works, he was thirty-five years old, and he brought with him enough papers, diagrams, and engineering drawings to fill six German army trunks. He found that Buz and Vogel had been equally industrious. A concrete foundation for the test motor had been poured in a brick workshop lit by high arched windows. The engine's cast iron A-frame, painted no-nonsense black, was already bolted down to its bed girders, and the single narrow cylinder was mounted on top of it. Most of the other parts were either in manufacture or being tested. Long experience with developing machinery had led Buz to install overhead lifting tackle and an auxiliary engine with a system of belts and clutches in the laboratory.

By July 25 the various parts had been assembled into a recognizable if odd-looking engine, with a bore and stroke of 150 × 400 millimeters, and a height of nearly 3 meters. Any doubts about the originality of Diesel's design would have been dispelled by an examination of the test machine. It was full of differences from previous prime movers and contemporary engines. Some of its features, like the long-ringless "plunger piston" sealed with a thin bronze band, were to prove impractical; others, like the high-pressure fuel injector and the combustion chamber in the piston crown, would become permanent features of later engines. There was a familiar-looking crankshaft, flywheel, and connecting rod, but the thicket of protrusions sprouting from the cylinder head was another matter. On an area only 190 × 300mm were clustered a dual intake-exhaust valve, an air starting and charging valve, a fuel injection needle valve, and an adjustable safety valve. As on all rationally tested engines, provision also had to be made for a Corliss or similar indicator. This small but invaluable device plotted the pressure versus volume conditions in the cylinder on a sheet of paper while the engine was running. It was one of the engine designer's most important diagnostic tools until the invention of electronic analyzers in the twentieth century.

The first two weeks of testing were devoted to lubrication and pressure measurements, with the engine driven externally. Every pump and pipe fitting gave trouble at the test pressures. One by one, each gasket and valve seat had to be removed, redesigned, made in other materials, replaced, and tested again. Leather, asbestos, and soft copper seals were juggled to find combinations that wouldn't leak. The commercial petcocks used as shutoffs in the air and fuel lines were hopelessly inadequate, and new ones had to be built to original designs. It was Newcomen's story again, but at a much higher level of precision. What was most disheartening was that these problems were occurring at pressures lower than the intended ones. After three weeks of troubleshooting, the cylinder pressure had been raised from 18 to 34 atmospheres, still well below Diesel's minimum specification. When every leak had been eliminated the trouble was found in the piston itself: the bowl-shaped combustion chamber had been cast inaccurately, so that its volume was about 60 percent larger than the designed figure.

Replacing the piston was a major project, and the rest of the engine now operated smoothly at 300 revolutions per minute. Encouraged by Vogel, Diesel decided to attempt a first combustion test on August 10 despite the nominally inadequate compression ratio. The indicator was set at zero, the piston driven up to the compression position with external power, and a charge of gasoline was sprayed into the cylinder. Ignition was instantaneous, the indicator pen shot up to 80 atmospheres, and the indicator and its valve exploded. The blast of flying glass and metal just missed Diesel and

Vogel, and they knew that at least one part of Diesel's theory worked in practice.

It was a tribute to Diesel's design that an unplanned explosion in a brand-new machine did no internal damage, but the spectacular one-shot ignition was not much different than Huygens' gunpowder engine, and nearly as far from continuous operation. The next day the engine was turned over by its auxiliary for many cycles, and fuel was injected regularly. In a short while erratic bangs interrupted the regular rhythm, and the laboratory was carpeted with thick black soot from the exhaust valve. The anemic pressure-volume diagram indicated only 1.58 kilowatts—not enough power to overcome the friction of the moving parts. After thirty-eight days of intensive testing it was clear that while compression ignition worked, sealing, cooling, combustion, and fuel injection were all unsatisfactory. The engine would have to be redesigned.

Rudolf was sobered by the realization that he had a long battle ahead of him. He returned to Berlin bringing Johannes Nadrowski, an assisting design engineer with him, and moved his family to a more modest apartment in the suburb of Charlottenburg. One room was equipped with a drafting table, and over the next four months new engineering drawings went from Charlottenburg to Augsburg on a regular basis. That Christmas and New Year were probably the most enjoyable ones the Diesels ever had. Rudolf rediscovered his family and took time out from engine design to build rabbit hutches, a playroom, and a Chinese shadow theater. When he returned to Augsburg, it was to a welcome extension of family life, and an echo of his own childhood.

About the time that Rudolf married Martha, Christoph Barnickel began courting Rudolf's younger sister Emma. Barnickel had been widowed for three years and was in his fifties, Emma in her twenties. They made a September-May marriage that according to all contemporaries was blissfully happy. Rudolf, for one, was delighted to have his old teacher and favorite cousin become his brother-in-law. In 1893 Christoph and Emma built a house over one of Augsburg's old canals and invited Rudolf to stay with them, and that was where he took up temporary residence in January, 1894.

Buz and Vogel had been as diligent as before, and two expert mechanics, Hans Linder and Friedrich Schmucker, were also assigned to the project. The much-modified "1894 engine" was almost ready to test when Diesel arrived. It now had a piston with sealing rings and a replaceable combustion chamber, separate intake and exhaust valves, a new air-blast fuel injector, and water cooling jackets around the cylinder and exhaust passage. All of the redesigned parts worked well except the most important ones: the fuel injectors.

It is no slight to Rudolf Diesel to say that fuel injectors have been the most troublesome components of the Diesel engine since it was invented. The requirements are frightful: a minute, precisely metered quantity of fuel must be delivered to the combustion chamber in a few thousandths of a second, at a time when the pressure in the cylinder is about 70 kilograms per square centimeter. There was no device in existence that could satisfy those requirements in 1894, and Diesel's struggle to make one can help correct a popular misconception about the invention of machines. Looking back, it is convenient to lump a set of changes together under the heading of a new design: "the 1894 engine." In real life it seldom works that way. Each improvement of a new machine is part of a continuous process of redesign, substitute, and test, over and over again. When a design "succeeds" it is simply another compromise, a decision that time, money, energy, or the urge to perfect has run out for the time being, and that things should remain as is. The success is (almost always) one of small triumphs over many tedious disappointments. It is impossible to know the exact number of individual changes that Diesel made in the fuel injection system of his engine before it was revealed to the public, but it must have been close to a thousand.

The first changes in 1894 were maddening, because they worked sometimes but not all the time. Erratic failures are loathed by engineers; it is much easier to fix a dependable flaw. All through January and half of February Diesel and his crew worked on the engine. The new parts were run in, and an external air compressor was added to the fuel injection system, which Diesel hated. Sometimes the injectors worked, sometimes they didn't; mostly they didn't. Every day the auxiliary motor turned the test engine at a few hundred revolutions per minute while fuel was injected and measurements made. On February 17, 1894, in the middle of a test run like any other, Hans Linder noticed the belt connecting the auxiliary drive to the flywheel of the test engine suddenly begin to jerk repeatedly—the Diesel was running the auxiliary! Linder raised his cap in a silent salute; Diesel turned and shook his hand, too moved to speak.

The engine ran for one minute: 88 revolutions on its own power. Diesel went home that night and wrote to Martha, asking her to hurry to Augsburg. She did, and when she lifted the starting lever her oily rival dutifully started and ran for her. One month later the German financial columns showed that Maschinenfabrik Augsburg's stock had risen 30 percent in thirty days of unusually lively trading. The day of the Diesel engine was coming, the market said. Like many rosy stock futures, it was a false dawn.

Eleven

One swallow doesn't make a summer. The proverb is the same in German as in English, and one minute's running doesn't make a successful engine. But common sense was no more common in 1894 than it is now, and while the hard-working mechanics were finding that they couldn't make the engine run smoothly or predictably, the stock market climbed and Diesel was invited to France and Belgium by interested licensees. In April of 1894 he began negotiations with Frédéric Dyckhoff, an old engineering friend from Paris who was a partner of F. Dyckhoff Fils in Bar-le-Duc. A few days later Carels Frères of Gent made a first payment of 20,000 francs against the rights to manufacture and sell Diesel engines in Belgium.

When Diesel returned to the factory, he found his crew still trying to coax a consistent performance from the test engine. It was more art than science; each control had to be watched and adjusted continuously. One record-breaking run of 36 minutes allowed a power measurement to be made: 9.7 kilowatts at 300 r.p.m. That was certainly better than 1.58 kilowatts, but it was accompanied by delayed ignition, violent detonations, and clouds of unburned oil from the exhaust port. Worse, there was no pattern to anything. Fuel injection was still the problem.

The next six months became an ordeal by experiment. Each change meant new drawings, hurriedly machined parts, modifications to the engine, reassembly, and more tests. It isn't surprising that Diesel became harried under the strain of continuous work and disappointment. The search took on a desperate quality as alternative after alternative was tried and rejected. Virtually every known hydrocarbon fuel was tested, from illuminating gas to kerosene, from high-octane gasoline to murky oils that had to be heated to make them flow. Fuel was injected prewarmed and precooled, in a stream, in a vapor, through screens, nozzles, and hollow needles, from direct plungers and remote pumps.

It is a measure of Diesel's frustration that he resolved more than once during this purgatory to give up his cherished principle of compression ignition and rely on auxiliary methods. The cylinder head was fitted with several kinds of internal wicks, insulated points energized by a magneto, and finally an electric spark igniter designed by the redoubtable Robert Bosch. But the system that operated perfectly in an Otto engine at a compression ratio of 2.5 to 1 utterly refused to work in the high-pressure caldron of a Diesel engine.

And yet, almost invisibly to those close to the machine, there was progress. The mechanics could have assembled the engine in their sleep by this time, and little by little their attention to details was yielding more reliability. Suddenly they were faced with a deadline. Krupp had applied for an Austrian patent to protect its sales rights. The Wien patent office required that a patent based on theory had to be validated by practical demonstration within a prescribed time to remain in force. The time limit for the Diesel patent was almost up. Of course there was only one possible engine to demonstrate. It was disassembled, shipped to the Krupp works at Berndorf, fitted with definitive versions of the improved parts, sent to Wien, and reerected for the Austrian patent commission. On January 17, 1895, fuel was supplied to the engine, and it ran. The patent was granted.

The Austrian test indicated that there were enough changes to warrant a major redesign of the prototype. The revised machine used the same frame and crankshaft, so its stroke remained at 400 millimeters, but the cylinder bore was raised to 220 millimeters. It was completed on March 26, 1895, and the familiar cycle of test and refit began again. Three months later to the day Diesel recorded the first dynamometer readings: Power: 16.93 kilowatts at 200 r.p.m.; Brake thermal efficiency 16.6 percent. While this wasn't a breakthrough in prime mover performance, it wasn't disgraceful either. As the inventor pointed out in a letter to the directors of Krupp and Augsburg, only a few engines could equal the efficiency figure, and the Diesel was still an experimental machine. Buz never wavered, and Krupp also agreed to continue with intensive development. Buttressing their decision, United States patent No. 542,846 for the Diesel engine, which had been pending for three years, was granted on July 16, 1895.

Diesel had resented the use of an external compressor for fuel injection since the early days of testing. He now began a two-pronged attack on the whole high-pressure problem. First, he accelerated work on a compound engine that he had proposed earlier, one in which the expansion would be divided into separate high- and low-pressure cylinders. Considering the difficulties he was having with a single cylinder, it was a rash step. Work on the compound engine continued over the next two years, with such disappointing results that the project was eventually abandoned. Second, he set himself the task of designing an integral air pump for the one-cylinder engine. The pump went through an evolution like the engine itself, requiring higher precision and harder steels. Meanwhile in the main engine new atomizing jets with holes in the pattern of a double star improved the combustion, and new lubrication passages raised the mechanical efficiency from 54 to 67 percent. After six more months of testing the engine could run for 50 hours before the fuel jets needed replacement, and its thermal efficiency was up to 20.26 percent.

On February 20, 1896, the backers held a major policy meeting and decided to concentrate on producing a marketable single-cylinder Diesel engine of 250 millimeter bore and 400 millimeter stroke. Diesel was to supervise both the drawing office and the assembly shop personally. A young engineer named Immanuel Lauster was hired to assist in the design of the production engine, and soon made himself indispensable. Even though the Krupp and Augsburg directors had recommended that testing of the laboratory engine be suspended in favor of work on the new model, it went on nonetheless. By the time the drawings for the new engine were finished on April 30, the old one was running continuously every working day from morning to night.

There were so many changes in the 1896 engine that another patent was applied for and received to cover them. Diesel moved his family to München so they could combine the life-style of a metropolitan center, now much to Martha's taste, with proximity to Augsburg. Every part of the new engine was tested at high pressure for casting flaws, and all lines and tanks were proofed against rupture. (It is indicative of Diesel's care that although many safety valves blew out in the Augsburg workshop, there was not a single injury during the entire five years of testing.) On September 7 the old engine was photographed for posterity and taken down, and one month later the new one stood in its place.

This machine looked very different from its ancestor of 1893. A cast pedestal carrying a crosshead guide for the piston rod formed half of its frame; the other leg was a tubular steel column which allowed easy access to the running gear. A row of lubrication lines marched across the front of the cylinder like a miniature pipe organ, and most of the visible valves and fittings had evolved into new patterns. The internal changes were even more drastic, including a water-cooled four-ring piston and (of course) an improved fuel injection system. The engine ran for the first time on December 31, 1896, and in late January it was deemed ready for testing.

The preliminary results were sensational. The first figure for brake thermal efficiency was 24.2 percent, significantly higher than the best gas engines available. The engine was not even broken in, and it was burning cheap kerosene, not expensive illuminating gas. Professor Moritz Schröter was summoned from München to direct a series of tests whose authority and objectivity could not be questioned. In the test made on February 17, the engine produced 13.1 kilowatts at only 154 r.p.m., with the unprecedented thermal efficiency of 26.2 percent at full load. What was even more important, it was not a fluke performance. The efficiency continued to climb as the engine ran in and the mechanics became more familiar with it.

Within a few days a parade of backers, engineers, and prospective licensees began to flow through the laboratory. Frédéric Dyckhoff came from Bar-le-Duc, and a group of thoughtful Scottish engineers representing Mirrlees, Watson & Yaryan of Glasgow put the new engine through their own rigorous tests. They were followed by Gisbert Gillhausen, director of Krupp, and two distinguished visitors, Director Schumm and Engineer Stein of Gasmotorenfabrik Deutz. Deutz was now the largest manufacturer of internal combustion engines in Europe, and Diesel's motor was made to run a tough gamut for them. It was started cold and immediately put under maximum load, accelerated to maximum revolutions, deprived of fuel, and started again immediately. Even under these circumstances it continued to put out more power and use less fuel than similar-sized I-C engines running under ideal conditions.

The reaction from Deutz leaves little doubt that its representatives went back to Köln frightened to death. The company informed the Augsburg-Krupp combine that they were deeply impressed by the new engine, but that they believed Diesel's patents could be questioned. They offered a token 5 percent royalty for the right to manufacture the Diesel engine, along with the implied threat of litigation. The directors of Krupp were shaken by Deutz's tactics and half-decided to withdraw their backing. Diesel, incensed at what he considered a slur against his honesty and originality, refused to even reply to Deutz's allegations. In a stormy eight-hour meeting in March 1897, he and Heinrich Buz convinced the faltering directors of Krupp to stiffen their backs.

Meanwhile Diesel himself was invited to London and

Glasgow to negotiate a licensing agreement. Perhaps it was characteristic that the Scots should have been quick to recognize the value of a prime mover that burned less fuel per kilowatt-hour than any other engine in the world. In a series of closely argued meetings, they drove as hard a bargain as legend would have led Rudolf to expect. Mirrlees, Watson's technical adviser was no less than William Thomson, Lord Kelvin, and Diesel was awed and honored to meet him. The two men hit it off quickly, and on March 23 1897, on Lord Kelvin's personal recommendation, Mirrlees, Watson signed the first contract to build Diesel engines outside of Germany. By the terms of this agreement Diesel would receive 20,000 marks annually. Until he visited the homes of Mirrlees and Robertson, two of the men who signed the contract, he might have regarded the payment as generous. He was astonished by the electrically-lit palatial houses of the Scots industrialists, and his envy continued to rankle after he returned to Germany.

On the way home, he stopped at Winterthur to visit Gebrüder Sulzer. He was received by Jakob Sulzer-Imhoof, later to be the head of the new Diesel engine department, and a tentative agreement was made in amicable and familiar surroundings. (Sulzers started building their first Diesel on August 11, 1897; it ran on June 10, 1898.) Rudolf had come a long way from *blaue Monteur*.

France was waiting to follow Scotland and Switzerland when Diesel arrived back at Augsburg. In April the Société Française des Moteurs Diesel was founded in Bar-le-Duc; Diesel visited the backers of the new company in June, and was voted stock with a value of 600,000 francs. On June 16 he had an honor that gratified him even more than financial success: he gave an invited address to the assembled Union of German Engineers at Kassel. A photograph taken that day shows an impressively dapper, top-hatted and white-tied Rudolf Diesel, looking almost shyly away from the camera. He stands arm-in-arm with Heinrich Buz, who dominates the picture like a kindly provincial patriarch; on Buz's other arm Professor Moritz Schröter twinkles at the photographer, obviously delighted to have believed in the new engine all along.

Diesel's talk was sober and scholarly, and emphasized the compromises of his engine more than its success. His experimental path was clear. Instead of improving an existing engine design, he had started from Sadi Carnot's thermodynamic principles and made them work. The Diesel engine began its power stroke with the working fluid at a higher temperature and pressure than any other engine. As Carnot had predicted, it extracted more power from its fuel than any other engine. It was also able to burn a wide range of fuels with higher energy contents to begin with than gasoline. The result was verified by Professor Schröter: Diesel had achieved his goal of producing the most efficient prime mover ever known. There was good reason why his speech should have been received with thunderous applause.

Despite this procession of triumphs, the directors of Krupp had continued to worry about the legal threat from Deutz. On July 19 Deutz demonstrated (1) that Diesel had been right to defy them, and (2) just how deeply impressed they had been with his engine. They signed a contract with the Augsburg-Krupp consortium for a 20 to 30 percent royalty on each Diesel engine they built, depending on size, and an immediate payment of 50,000 marks to Rudolf Diesel. A few weeks later Diesel bought the most expensive site available for a villa in München, on Maria-Theresia-Strasse overlooking the Isar River. The price of the bare lot was 50,000 marks.

Rudolf Christian Karl Diesel, inventor of the Diesel engine.
Born in Paris, France, March 18, 1858; died in the
English Channel, September 29, 1913.

Rudolf Diesel in 1883, at age twenty-five.

Martha Flasche Diesel as a young woman. She was born in 1860 and died in 1944; this picture was probably taken near the time of her marriage to Rudolf Diesel in 1883.

Some of Rudolf Diesel's preliminary sketches for the "rational heat engine."

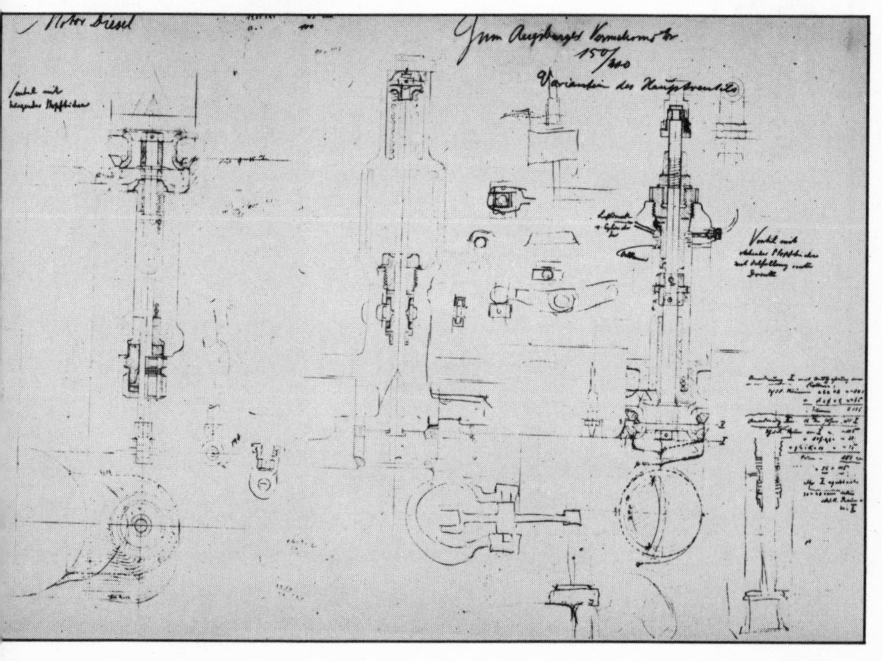

A model diorama of the workshop at Maschinenfabrik Augsburg where the first experimental Diesel engines were built and tested. The engine shown is the earliest version, assembled in July 1893.

The patent certificate issued by the Imperial German Patent Office to Rudolf Diesel for the development of his internal combustion engine.

PATENT-URKUNDE

№ 67207

AUF GRUND DER ANGEHEFTETEN BESCHREIBUNG UND ZEICHNUNG IST DURCH BESCHLUSS DES KAISERLICHEN PATENTAMTES

an Rudolf Diesel, Ingenieur, in Berlin

EIN PATENT ERTHEILT WORDEN.

GEGENSTAND DES PATENTES IST:

GESETZ v. 7. APRIL 1891

Arbeitsverfahren und Ausführungsart für Verbrennungskraftmaschinen.

ANFANG DES PATENTES: 28. Februar 1892.

DIE RECHTE UND PFLICHTEN DES PATENTINHABERS SIND DURCH DAS PATENTGESETZ VOM 7. APRIL 1891 (REICHS-GESETZBLATT FÜR 1891 SEITE 79) BESTIMMT.

ZU URKUND DER ERTHEILUNG DES PATENTES IST DIESE AUSFERTIGUNG ERFOLGT.

Berlin, den 23. Februar 1893.

KAISERLICHES PATENTAMT.

Beglaubigt durch

Bureau-Vorsteher des Kaiserlichen Patentamtes.

An accurate (if slightly idealized) depiction of the first Diesel engine. It had a bore of 150mm, a stroke of 400mm, and stood nearly 3 meters tall. It developed only 1.58 kilowatts—not enough to overcome the friction of its own moving parts—and was redesigned after a few weeks of testing.

The rebuilt test engine, which went through hundreds of modifications between 1893 and 1896. This version had its bore enlarged to 220mm, and was the first Diesel engine to run under its own power: 1 minute at 88 r.p.m., on February 17, 1894.

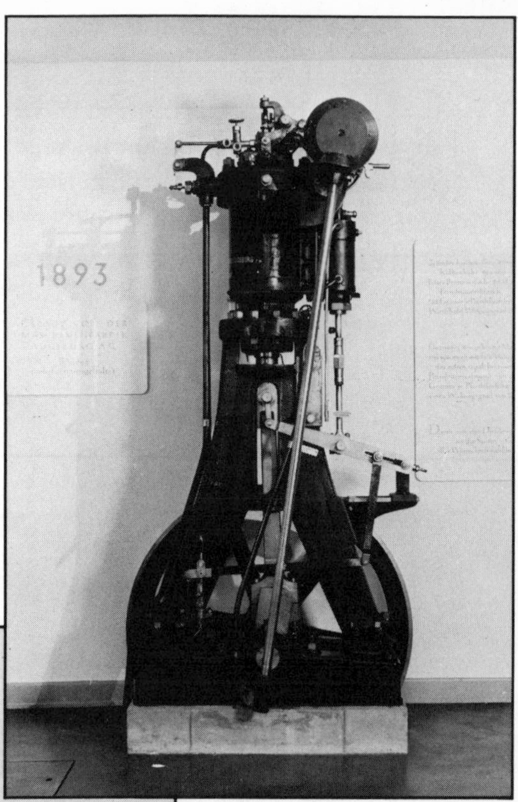

The machine that realized Diesel's ambition to build the most efficient heat engine in the world. Bore, 250mm; stroke 400mm. When proof-tested by Professor Moritz Schröter of the München Technische Hochschule on February 17, 1897, it developed 13.1 kilowatts at the unprecedented thermal efficiency of 26.2 percent.

The Kollektiv-Ausstellung von Diesel-Motoren at the München Power and Works Exhibition of 1898. Diesel powerplants built by four German manufacturers were exhibited side by side in this pavilion, and drew thousands of visitors.

A day of triumph. Rudolf Diesel with Heinrich Buz, Managing Director of Maschinenfabrik Augsburg, and Professor Moritz Schröter at Kassel on June 16, 1897, where Diesel gave an invited address on his engine to the Union of German Engineers.

Kiev, Russia, 1904: The first Diesel-electric power station in the world. Six M.A.N. engines produced 1800 kilowatts to power the streetcar system of the Kiev Municipal Transport Authority.

Prosper L'Orange (1876–1939). This talented engineer born in Beirut, Lebanon, invented the precombustion chamber and later perfected the airless fuel injection system that made the Diesel engine a practical powerplant for motor vehicles.

A double first. This is the first two-stroke Diesel and the first directly reversible Diesel. It was built by Gebrüder Sulzer of Winterthur, Switzerland, in 1905 and had an output of 66 kilowatts.

The Norwegian topsail schooner *Fram*, a ship that deserves a biography of its own. Built for Fridtjof Nansen in 1892, she was intentionally frozen into the arctic icepack for three years, drifting to 86° North latitude. In 1910 she was fitted with a Swedish Diesel engine and carried Roald Amundsen to Antarctica in his successful bid to be the first man to reach the South Pole (December 14, 1911).

The *Fram*'s 132-kilowatt engine, built by AB Diesels Motorer of Sickla, Sweden. When Amundsen reached Tasmania in March 1912 he telegraphed the makers, "Dieselmotor excellent," and they promptly adopted the engine trademark "Polar."

Ivar Knudsen, the technical director of Burmeister & Wain of København, Denmark, who persuaded his company to acquire the manufacturing rights for Diesel engines in 1898. He directed the construction of *Selandia*, the first ocean-going motorship in the world.

Winston Churchill, First Lord of the Admiralty, tours *Selandia* during her visit to London in 1912. Accompanying him are Ivar Knudsen of Burmeister & Wain, and H. N. Anderson, Director of the Danish East Asiatic Company for whom the ship was built. After his inspection Churchill called the ship "... the most perfect maritime masterpiece of the century."

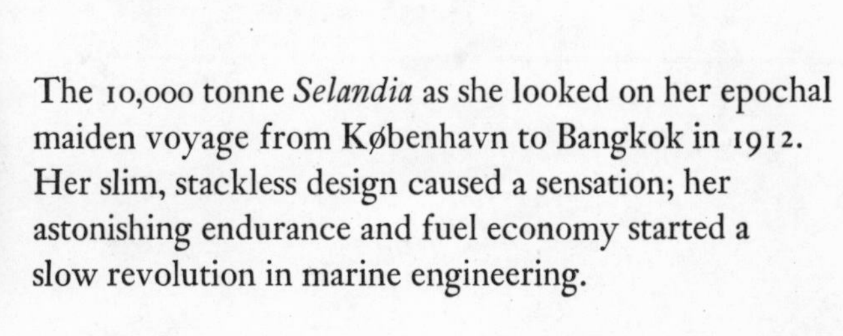

The 10,000 tonne *Selandia* as she looked on her epochal maiden voyage from København to Bangkok in 1912. Her slim, stackless design caused a sensation; her astonishing endurance and fuel economy started a slow revolution in marine engineering.

The first Diesel locomotive. This 80-tonne joint project of Sulzers, Krupp, and the Prussian & Saxon State Railways was completed in 1912, but was plagued with transmission problems.

Sweden's Mellersta & Södermanslands Railway commissioned this Diesel-electric railcar from ASEA in 1913. Powered by an AB Diesels Motorer engine, it was the first revenue-earning Diesel railway vehicle and paid its way until 1939.

In 1914 Sulzer converted its locomotive experiment to profit with a fleet of five of these 147-kilowatt Diesel railcars for the patient Prussian & Saxon Railways.

By 1917 the pressures of war had forced the development of very sophisticated Diesels. This 10-cylinder M.A.N. U-Boat engine used high-pressure air to inject its fuel and developed 2200 kilowatts at 300 r.p.m.

The Benz 5K3. Introduced in 1923, Benz's 5-tonne truck marked the debut of the direct fuel-injection Diesel engine in commercial road vehicles.

M.A.N.'s first direct-injection Diesel truck, marketed in 1924, was competition for Benz, but still retained compressed air for its horn.

The first mobile Diesel to succeed in the United States. The Caterpillar 1C1 Diesel-engined crawler tractor was brought out in 1931 and outsold competing gasoline-engined models by the thousands through the depths of the depression.

While Caterpillar crawled to success, Clessie Cummins of the Cummins Diesel Company tried to convince a conservative America that the Diesel could run fast as well. Here is an early 1930s transport of a Cummins Model U engine into an impressively robust automobile chassis.

Only one cylinder, but what a cylinder!
The research engine built by Sulzer Brothers in 1924
had a piston 900mm in diameter. It developed 1765 kilowatts
at a leisurely 120 revolutions per minute.

Twelve

There was, to be sure, Martha's determined ambition to live in a mansion. But even allowing for that, Rudolf's purchase of the lot in München was reckless. Unfortunately it was also typical of his financial decisions. It was strange: this rational engineer, a near-teetotaller, modest in his behavior and obsessed with the ideals of economy and practicality in his inventions, had a complete blind spot where money was concerned. Perhaps it ought to be called a perverse talent; no matter how much money he made, it ran through his hands a little faster than he could replace it.

In July 1897 the Diesels moved to a spacious apartment at Schackstrasse 2. The household now included a butler and a French governess for the children, as well as the usual below-stairs servants. Rudolf and Martha went often to theaters and the opera, where the works of Richard Wagner had become Rudolf's overwhelming favorites. Martha also began to have many social activities that didn't overlap with her husband's. Rudolf took up photography as a hobby, and an architect was hired to begin drawings for a house. Diesel commented to a new friend, Graf Ferdinand von Zeppelin, that he planned to build a modest home on an appropriate site. It is clear from what he did build that his real aim was to emulate the great mansions he had visited in Scotland.

In September of 1897 Diesel was contacted by the representatives of a man to whom Mirrlees' and Roberston's homes would have seemed mere cottages. This was Adolphus Busch, the brewery magnate of St. Louis, Missouri. The Busch family had emigrated to the United States from Rheinhessen and become enormously rich by making excellent beer. But Adolphus Busch was a farsighted industrialist as well as a brewer; when a German hops dealer told him about the new engine being built in Augsburg, he sent his technical adviser,

Colonel Edward Meier of the U.S. Army Engineers, to have a look at it. Meier arrived in Augsburg accompanied by George Marx, the chief engineer of another distinguished German company, *Maschinenbaugesellschaft Nürnberg.* Meier and Marx received carte blanche treatment from Heinrich Buz. While they filled notebooks on the Diesel engine and its inventor and builders, Herr Busch himself traveled from America to the spa of Baden-Baden with fifty friends, relatives, and staff.

Meier reported to Busch personally on October 4, as enthusiastic about the possibilities of the Diesel engine as the licensees who had already committed themselves to it. Several days later Rudolf Diesel received a cordial invitation to visit Adolphus Busch at Baden-Baden. Busch's reputation must have preceded him, but even with forewarning it was hard not to be impressed. His suite occupied an entire floor of one of the most luxurious hotels in Germany. The man himself was jovial and approachable, but his suit pockets swelled with the only form of money that he carried: solid gold. For a small service he tipped a small gold piece, for a larger service a larger gold piece. There seemed to be no shortage of people eager to help him.

Colonel Meier had strongly recommended that Busch acquire the rights to manufacture and sell Diesel's engine in the United States. After some cordial interaction, Busch asked Diesel how much an exclusive American franchise for the Diesel engine would cost. Diesel must have taken a breath before naming a sum that was worthy of Busch's position as a prince of industry: one million marks. Busch was courtly and matter-of-fact; he called for a checkbook, wrote a check, and asked Diesel to have the contract drawn up for signing. On October 9, in München, both men signed the contract, and Busch gave Diesel the same check, for 1,000,000 marks payable in gold. It was worth $238,000 at the prevailing rate of exchange, and at a guess its real purchasing power would have been about $800,000.

Three other major industrial firms sent representatives to Augsburg in the same month as Busch: Vickers Sons & Maxim, Ltd. from England, *Gebrüder Howaldt-Werft* of Kiel, and *Aktieselskabet Burmeister & Wain's Maskin og Skibsbyggeri* of København, Denmark. Diesel was especially pleased to be contacted by the latter two shipbuilding companies. In his manuscript of 1892 he had written, "It is almost impossible to estimate the advantages of using smaller engine units [in ships] . . . and of the abolition of boilers." The Danish engineering firm of Burmeister & Wain was destined to demonstrate those advantages dramatically a few years later.

The flood of licensees was very welcome to the managers of Augsburg and Krupp. The two firms had thus far spent more than 600,000 marks and five years on the development of the Diesel engine. The promise of recovering their investment through sales and royalties kept work on the engine going at a rapid pace. The Augsburg design team changed somewhat under the pressure. Lucian Vogel, disaffected by the drive to get a Diesel engine on the market without more testing, resigned from Maschinenfabrik Augsburg late in 1897. Diesel was sorry to lose him. He later wrote that "Lucian Vogel supported my concept with an inner conviction from the start, and never during the long and difficult search for success did he show a moment of doubt or vacillation. He gave advice, hard work, and the benefits of rich experience unselfishly and unfalteringly, and contributed many valuable ideas to the project as well."

Immanuel Lauster now began to come into his own as an engine designer. Lauster directed the design of the first twin-cylinder Diesel, and the result was more than a mere coupling of two single-cylinder engines. Bore and stroke were both enlarged from the 1897 prototype, and many refinements resulted in a much smoother-running machine. The rated output—conservative, as usual for Augsburg's products—was 44 kilowatts at 180 r.p.m. The new twin was just finished as 1897 drew to a close.

The New Year came in like a lion. On the first day of 1898 Adolphus Busch's Diesel Motor Company of America opened its offices at 11 Broadway in New York. On the same day in Augsburg two German bankers chartered another new Diesel engine factory. *Die Dieselmotorenfabrik Augsburg* was capitalized at 1,500,000 marks, 100,000 of which were Rudolf Diesel's. Diesel was a director of the company, and found himself in the odd position of authorizing the purchase of a license from Krupps and Maschinenfabrik Augsburg to build his own invention. He wasn't comfortable about it, but it seemed like only a small rift in the tide of good fortune. A few weeks later, two Swedish financiers, Marcus Wallenberg and Oscar Lamm, bought the rights for the manufacture and sales of Diesel engines in Sweden and founded still another new company, AB Diesels Motorer, to build them.

But probably the best news for Rudolf Diesel in January 1898 was the sale of the first industrial Diesel engine. The customer was *Aktiengesellschaft Union*, a match manufacturing company located in Kempten, a town about 80 kilometers southwest of Augsburg. The engine was Lauster's new twin; it was delivered in January and started on March 5. Perhaps this is the place for a note on the size and power of these early stationary engines. Forty-four kilowatts doesn't sound like much to people conditioned by advertising claims of hundreds of horsepower from automobile engines. Besides being measured under idealized conditions not likely to occur in service, that horsepower can only be produced at high speed. The Kempten Diesel had pistons the diameter of a long-playing record, and they moved very slowly, generating tremendous torque. The engine stood taller than a man, and its final drive was taken from two massive flywheels of the same order of diameter. That one engine could power most of the machinery in a medium-sized factory. Incidentally, the low operating speeds meant that the engine lasted a long time. Fifteen years later the first commercial Diesel was still running perfectly and had never needed a major repair. The experts who inspected it concluded that it would give many more years of trouble-free service.

Rudolf Diesel would have been grateful for the same assurance at the time the Kempten engine was installed. He had all the outward signs of success, but he felt worse and worse. The headaches were back, more frequent and excruciating than ever, and he began to suffer from twinges of gout in his right leg and foot. Though he confided to a friend, Ludwig Noé, that he felt very sick and couldn't bear the pains in his head, he mastered his internal torments repeatedly during the financial meetings that continued to occupy his time.

On February 16, 1898, he signed a license agreement with another world-famous industrial family. Emanuel Nobel, the Swedish-Russian nephew of Alfred Nobel, sent his engineers to Augsburg, and they reported back with the same glowing recommendations as others before them. Nobel promptly founded the Russian Diesel Company of Nürnberg, with the right to manufacture and sell Diesel engines in all the Russias. In exchange for those rights Rudolf Diesel received 600,000 marks in cash, and 200,000 marks in the stock of Nobel's company.

One would have expected Diesel to view the approach of his fortieth birthday with equanimity and satisfaction. His engine had fulfilled his prediction, and companies were being formed to build it in nearly every industrial country. He was still very handsome and bore his age with distinction. His three children were healthy, intelligent, and good-looking, and his wife traveled in the long-desired social circles. He himself was a millionaire, on paper at any rate, five times over, and he was planning to build a house that would make that clear even to someone who had never heard the name Rudolf Diesel. As a capstone the patent lawsuit brought against him by the French inventor Emil Capitaine, which had cost him much agonizing worry, was decided in Diesel's favor on April 21.

Three months later the München Power and Works Machinery Exhibition opened, and its feature exhibit was the *Kollektiv-Ausstellung von Diesel-Motoren*. Thousands of visitors came to the custom-built pavilion to see four big Diesel powerplants running side by side. Augsburg, Krupp, Deutz, and Nürnberg-built engines operating simultaneously produced exhaust that was almost invisible against the smoke from a single steam plant nearby. Tourists gaped as attendants demonstrated the strange properties of air liquefied by a Diesel-driven Linde compressor, and Brakeman and Sulzer high-pressure pumps powered by the Augsburg and Krupp engines threw mighty jets of water out into the Isar River. Not long afterward the Augsburg and Nürnberg factories decided that their collaboration during the exhibit should be extended. The two firms merged in the fall of 1898, under a title that even Bavarians found too long. In 1904 it was shortened to the name by which the company has been known ever since: *Maschinenfabrik Augsburg-Nürnberg AG*, or M.A.N.

At about the time of the merger Diesel realized that it was impossible for him to keep administrating the expansion of Diesel engine licenses singlehanded. A group of eminent backers was convened, and a company formed that henceforth would manage the development and improvement of the Diesel engine, as well as the sale of manufacturing and vending rights in foreign countries. Its equivalent name in English would be General Diesel Corporation. The company contracted to buy from Rudolf Diesel all patents, rights, and records that he had thus far acquired for the Diesel engine, in return for a payment of 3,500,000 marks in cash. The company was officially founded on September 17, 1898, and in the event Diesel received 1,250,000 marks in cash and the rest in stock certificates.

A few weeks later the various doctors that Rudolf had been seeing agreed on a diagnosis of nervous exhaustion, and recommended that he take a rest cure at a private sanitarium in Neuwittelsbach, northwest of München. He entered the sanitarium in October, bringing with him copies of the house plans for the lot on Maria-Theresia-Strasse.

Thirteen

The rest cure was not a success. At first glance it was difficult to see that Diesel had much to worry about, and he himself tried to give that impression. He seldom complained except to Martha, and, on rare occasions, to a few close friends. It was as though he felt ashamed to admit that he had problems and disappointments like other people; he simply internalized them instead. As of 1898 his real disappointments were few. Professionally, he had to give up the compound engine project. In the long run that was a blessing; the unwieldy compound had high heat and friction losses, and it developed about half the output predicted for it. Personally, he worried about money. His reaction was to spend it so lavishly that no one—not even himself, perhaps—could imagine that he had financial problems.

At the end of the year he came home for the Christmas holidays, then returned to the sanitarium in January. Meanwhile the house overlooking the Isar was beginning to rise on a scale of modernity and magnificence that München had never seen before. Diesel's income at this point was fantastic, but the severity of his headaches increased in proportion to the bills. The doctors decided that Neuwittelsbach was too close to home. Diesel allowed them to send him to Schloss Labers, a rest-castle for very rich and mostly very old people in the Tyrolean Alps. It was a relief for him to return to München in April, even though the headaches still raged.

At Augsburg his erstwhile engine group had begun experiments on one of his pet projects: the attempt to burn coal dust in a Diesel engine. Diesel had mentioned this as a potentially important method of energy conversion in his address at Kassel, and had tried to convert Heinrich Buz to his views. Coal-burning tests were begun in 1899, and after a short time a Diesel engine was started on kerosene and switched over to coal dust operation. But the first run was only 5 minutes, the second 7 minutes. To the directors of M.A.N. it looked too much like 1893 again, and the project was aborted. (A successful coal-dust engine was eventually developed by Rudolf Palikowski, who had been an engineering assistant to Diesel in 1897).

Many modern corporations insist that their executives recuperate after an illness, or even a long airplane flight, before making any crucial decisions. Diesel could have used a proscription like that in 1899. While at Schloss Labers he invested heavily in a Balkan petroleum development. Late in the year, when the Diesel family was preparing to move into their new mansion, Rudolf heard that the venture was failing. He decided to go to Lvov (then in Austria-Hungary, now in the Ukraine) and inspect the oil fields himself. It must have been a disheartening trip; he returned with the news that he had lost about 300,000 marks on his stock.

The loss came at a particularly bad time. Diesel's bank accounts, which over the past two years had been swelled by many five- and six-figure deposits, were very low. Where could all that money have gone? Most of it went into building Maria-Theresia-Strasse 32. Even on a street of magnificent villas, Diesel's home was something special; it showed a rare combination of technical expertise and capital investment. The extra-deep foundations were air-spaced to keep out moisture. The basement had a bicycle corridor for the children to ride in on rainy days, and the house was plumbed and wired to a standard new for München. There were five bathrooms, and every principal room had marble-hearthed fireplaces. Even the double-opening windows were custom-designed.

The ingenuity expended on the structure of the house was not so evident in the furnishings. Perhaps it was Martha who insisted on standard *grand luxe:* vaulted, carved and painted ceilings, heavy French furniture and Italian wardrobes. The sofas in the conservatory were upholstered in silk, and Martha's rococo sanctorum was an ivory-walled, Persian-carpeted Louis XV salon, complete with a rarely played grand piano. Rudolf's second-floor study was fit for a designer-prince. It was the only room in the house decorated in what was considered "modern" at the time, with twining *Jugendstil* motifs, a brown marble fireplace, and a deep yellow oriental rug. Despite its generous wall space to hang drawings and a surfeit of drawers for papers, Diesel liked this palatial workroom less than any of its modest predecessors. Not until he had slowly transformed the study and several adjoining rooms into an approximation of the old Augsburg workshop did he really tolerate his luxurious surroundings.

The end of 1899 was a time of peculiar excitement in Europe and America. The numbering of calendar years is of course arbitrary, and differs from culture to culture, but it is hard to resist the feeling that a change of centuries or millennia heralds something significant. In Paris the greatest international exposition of all time was assembled to celebrate the turn of the century. It opened on April 14, 1900, and during the next eight months more than fifty million people came to see what had been accomplished, and what was likely to be accomplished. There are many products that still carry on their labels reproductions of the gold medals won at Paris in 1900, but the *Grand Prix*—the highest honor awarded to any exhibit—went to the Diesel engine.

It was a valid portent. The first decade of the twentieth century was to see the penetration of Diesel power into many areas hitherto monopolized by steam. But while the new engines succeeded (at a slower rate than predicted), the fortunes of the Diesel family fluctuated precariously. Luckily, not all the oscillations were downward, and there were some diversions as well as alarms.

On July 2, 1900, Rudolf was personally invited by Graf von Zeppelin to witness the first flight of the Count's new airship LZ.1 from Lake Konstanz. *Luftschiff Zeppelin 1* was 128 meters long and 11.7 meters in diameter. Her gas capacity was 11,300 cubic meters, and her loaded weight 13.1 tonnes.

This immense craft was powered by two 10.3-kilowatt Daimler gasoline engines, each driving a pair of propellers through outboard shafts, and the ship was to be steered by two diminutive rudders at the bow and stern. LZ.1 took off carrying five men (including the 62-year old Count), rose to a height of 300 meters, and made a one-hour, semicontrolled flight, eventually landing safely on the lake with various malfunctions. The unconventional Graf recognized in Diesel a man whose obsession matched his own, and asked him to develop a version of his engine that could power an airship. This was at a time when the lightest Diesel engine weighed about 270 kilograms per kilowatt! Rudolf thought the idea was feasible and made some design sketches, but the concept was decades ahead of its time.

By May 1902 there was a known total of 359 Diesel engines in use, with a combined output of 9,102 kilowatts. The maximum power per engine was rising rapidly, and licensees in many countries were making improvements to the basic design.

The following year M.A.N. began construction of a major Diesel-electric generating plant to be installed in Kiev, Russia. Although Emanuel Nobel held the Diesel manufacturing rights for Russia, and his St. Petersburg factory had built several 75-kilowatt pumping engines, he could not supply the larger units needed to power the streetcar system of the Kiev Municipal Transport Authority. M.A.N. guaranteed the entire installation, including the electric generators supplied by another contractor. The plant eventually comprised six four-cylinder Diesels, each with an output of 300 kilowatts at 160 r.p.m. The Kiev station was the largest Diesel powerplant in the world for some years, and proved very successful.

Nobel shared Diesel's optimism about marine propulsion. There was one problem: a ship had to be able to go astern, and no one had yet produced a Diesel motor that could be reversed like a steam engine. Nobel was determined to experiment anyway. He had a fleet of shallow-draft river tankers to carry oil from the fields at Baku on the Caspian Sea to St. Petersburg, a trip of more than 4,000 kilometers. In 1903 K. W. Hagelin, Nobel's engineer in charge of oil transport, began the construction of the first Diesel-propelled cargo vessel in the world.

Vandal was unique in many ways. She was 74.5 meters long by 9.68 meters in the beam, but she drew only 2.44 meters of water fully loaded. Her long deck trunk and tiny bridge house amidships gave her a profile remarkably like that of a World War II submarine. In common with other Nobel tankers she carried 700 tonnes of oil in four tanks, but there the similarities with her steam-driven predecessors ended. Her engine room housed 3 three-cylinder Diesel engines built by AB Diesels Motorer of Sickla, Sweden. Each engine developed 88 kilowatts at 240 r.p.m., and to solve the reversing problem, each one drove an 85-kilowatt electric generator, which in turn powered a 75-kilowatt reversible electric motor coupled to a propeller. The system was almost identical to the one used today in Diesel-electric locomotives.

The ship's equipment also included an auxiliary generator for winches, steering gear, and lighting, and two electric pumps that could unload her cargo in six hours. All engine functions were controlled directly from the bridge, and her triple screws could drive her against the Volga River's current at 22 kilometers per hour fully loaded. *Vandal* was an immediate success. In 1904 she delivered the same tonnage of oil in a similar time as earlier Nobel vessels, but she did it on only one-fifth the fuel. Nobel promptly had the chartered tanker *Ssarmat* equipped with two 132-kilowatt variable speed Diesels built in St. Petersburg, and made plans to convert his other ships.

During the same year the *Anciens Établissements Sautter-Harlé* of Paris, licensed by Diesel in 1899, built an 18-kilowatt opposed piston Diesel that would run in either direction without a gearbox, and installed it in the tiny Belgian canal boat *Petit Pierre*. Diesel himself took a ride on Little Peter and smilingly endorsed this new development. The next five engines that Sautter-Harlé built were much larger, but they did not go into canal boats; they went directly to the French Navy, which installed two of them in the experimental submarine Z, and the others in three more submarines.

It was the first inkling that the Diesel engine was going to find applications that its inventor, for all his foresight, had not foreseen. Diesel had great difficulty reconciling that fact with the desire for his engine to succeed. He considered himself an internationalist and a pacifist. When Americans asked him if he was German, he invariably corrected them with great emphasis: "No, Bavarian. My paternal ancestors were Slavs. I am French born and Bavarian naturalized; I like to be a citizen of the world." It was another blind spot, like money. Diesel's first reaction to the increase of international tensions was to turn away and immerse himself in other areas.

In 1904 he went to the Gordon Bennett race and saw France's Jacques Théry on a Richard-Brasier beat Camille Jenatsky's Mercedes at an average 87.7 kilometers per hour over 533 kilometers of German roads. Rudolf came home with an itch familiar to many race spectators and bought a bright red *NAG* touring car powered by an 18-kilowatt gasoline engine. His gouty right foot was not dependable, he found, and his vision was also less than perfect. The car had a nominal top speed of about 35 kilometers per hour, and Diesel hired a chauffeur to drive it most of the time.

As a distraction the car was a failure, but its owner soon found another one. In June 1904 he sailed on the liner *Pretoria* to New York. After several days in Manhattan, he traveled by train to St. Louis, the Louisiana Purchase Exposition, and a handsome welcome from Adolphus Busch. Busch was having great difficulties launching the new engine in America. He had begun by assembling European-built Diesels. His first U.S.-built engine, a three-cylinder 55-kilowatt unit, ran in April 1902, but at the time of Diesel's visit fewer than 100 engines had been sold, many of them without profit. Diesel's advice didn't make any immediate change in the fortunes of the American Diesel Company, but the trip was an eye-opener for him. During the next month he traveled the breadth of the United States. He recorded in copious notes his amazement at the size, the generosity, the waste, and the lack of culture. At the end of his trip he saw little future for the Diesel engine in America, because the country seemed to him to have an unlimited supply of cheap fuel and raw materials, and a population determined to use them up as fast as possible.

Somehow the American journey succeeded where Schloss Labers had failed. For almost a year Diesel felt better physically, and he turned his attention to the horseless carriage. Working with an assistant, Heinrich Deschamp, he designed a "petite model" Diesel engine that developed 3.7 kilowatts from four small cylinders. A larger 22-kilowatt model was installed in a van chassis, but its power was inadequate for the weight of the engine and air compressor. Although few were sold, the "petite" later won a Grand Prix at the Brussels World Fair of 1910. Diesel was still producing ideas ahead of their time.

As 1907 came to an end he faced the prospect of several momentous changes. First his daughter Hedy announced her forthcoming marriage to Arnold von Schmidt, a young engineering graduate of Diesel's own alma mater. The wedding was one of the more glittering events in München's social calendar for 1908. Diesel gave his daughter away stoically, but he was wracked with violent headaches for days afterward. He was also nearly half a million marks in debt, having lost about 3.5 million marks through unwise investments in the past five years. The other change was that the basic European patents on the Diesel engine expired.

Fourteen

The expiration of the first Diesel patents started a flood of engine building in many countries. Some of the most significant progress was made close to home. At Deutz a young engineer and admirer of Diesel named Prosper L'Orange began to design a new fuel injection system that required no external air compressor. Benz & Cie had an appropriately modern response to this modern idea: They enticed L'Orange away from Deutz and gave him his own laboratory. At Benz, L'Orange continued to work on his injection system and invented another important advance in Diesel technology: the *vorkammer* or precombustion chamber. This is a small, precisely shaped compartment in the cylinder head, which is connected via ports to the main combustion space. The fuel injector discharges into the precombustion chamber, and the fuel ignites there first and then expands into the cylinder. L'Orange found that vorkammer Diesels ran smoother and less noisily than cylinder-injection engines, though at a slight sacrifice in fuel economy. Benz & Cie patented the precombustion chamber in 1909, and with successive improvements it has been a feature of their Diesel engines ever since.

During this same period R. Hornsby and Sons of Grantham, England were marketing a sturdy low-compression oil engine invented in 1890 by Herbert Akroyd Stuart, a practical mechanic born in Yorkshire. Stuart's engine also used a chamber in the cylinder head as an ignition site, a cast iron "hot bulb" that had to be brought to red heat with a blowtorch or pilot flame before the engine would start. Once the engine was running the heat of combustion kept the bulb hot enough to ignite fuel sprayed into it.

Because of the superficial similarity of arrangement, some partisans insisted that Stuart had anticipated Diesel and L'Orange in inventing the vorkammer Diesel principle.

Stuart himself did not press this claim; in fact, after the Hornsby version of his engine was brought out in 1892 he rarely visited the factory again. His design had the virtues of simplicity and ruggedness, but it was neither a compression-ignition engine nor a "semi-Diesel" as several companies misleadingly named it. The Hornsby-Akroyd engine that won a medal at the Cambridge Royal Agricultural Show in 1894 took 8 minutes to start, developed 6.4 kilowatts at 240 r.p.m., and had a brake thermal efficiency of 14 percent. This was a production model that had been on the market for two years; its efficiency was lower than the first dynamometer test on the prototype Diesel, and less than half the full-load efficiency of the Augsburg test engine of 1897. At 318 kilograms per kilowatt it was even heavier than the early Diesels, but it was also cheaper and less difficult to build. For about twenty years it sold by the thousands, filling a need for agricultural and fishing boat engines that could be maintained without skilled mechanics.

Sulzers of Winterthur worked at the opposite end of the engineering scale. These citizens of a small, landlocked country pursued two surprising goals: marine powerplants and big engines. Their first waterborne Diesel, installed in 1904 in the Lake Léman freight boat *Venoge*, was much like *Vandal*'s and used a similar power train. Sulzers were dissatisfied with electric drive, and to solve the reversing problem they decided to develop a two-cycle Diesel.

There were of course other potential benefits besides reversing in a two-cycle engine. The basic argument for it was similar to the one advanced for the gasoline ignition engine in the 1870s: Why waste two strokes of every four using the piston as an air pump? The problems were also similar: Combustion time was limited; burned gases had to be swept out of the cylinder and a fresh charge brought in during a short interval at the end of the power stroke and the beginning of the compression stroke; the two-cycle required forced air for scavenging, while a four-stroke was self-aspirating.

Sulzers' engineers realized that despite these problems the Diesel offered some comparative advantages for two-stroke operation. Its development had been accompanied by the design of powerful air pumps, and scavenging a Diesel cylinder with a surplus of pure air was easier and more economical than using the mixture of air and fuel required by a gasoline engine. The Diesel was also a relatively slow-running machine, with a little more time to get the burned gases out and the fresh air in. A two-cycle engine nominally has twice the output of a four-stroke of the same displacement, so that half the number of cylinders can produce the same power. This is particularly useful in the Diesel because it reduces weight and the number of expensive fuel injection components.

It took the Swiss only one year to convert these ideas to practice. Their new marine engine, erected in 1905 and an object of much attention at the 1906 Milano World Exhibition, was a double first: The first two-cycle Diesel engine, and the first directly-reversible Diesel (as opposed to one which could only be started in either direction) of any kind. It developed 66 kilowatts from four cylinders of 175mm bore and 250mm stroke. Sulzer-Imhoof's next project was to test the two-stroke principle in sizes large enough to power ocean-going ships. In 1911 the Winterthur shops built a giant one-cylinder Diesel with a piston 1 meter in diameter. This monster, turning at a majestic 150 r.p.m., produced 1,472 kilowatts! Sulzer, encouraged, immediately began work on a four-cylinder version to produce 2,760 kilowatts.

While the Swiss pioneered mountainous Diesels, the maritime nations of northern Europe took up where Emanuel Nobel had pointed. In 1909 the Danish East Asiatic Company ordered from Burmeister & Wain a vessel that was to make maritime history: *Selandia*, the first ocean-going motorship in the world. The driving force behind *Selandia*'s construction was Ivar Knudsen, B & W's technical director. Knudsen had urged the company to acquire the Danish rights for the Diesel engine in 1898. Diesel later wrote that Knudsen was "... the man who in the whole world has best understood my ideas and been able to improve on them as well." Then as now B & W was a painstaking company, and they were determined to produce both the ship and her engines to the highest standard. *Selandia* was launched on November 4, 1911, but while she was still on the stocks, the Dutch were finishing a small version of the same idea. They installed a 338-kilowatt Amsterdam-built Werkspoor Diesel in the little tanker *Vulcanus*. *Vulcanus* displaced only 1,200 tonnes and was even shorter (59.6 meters) than *Vandal*, but the Dutch sailed her out to her intended station on the island of Borneo without a hitch. She gave excellent service there until she was broken up in 1931; her engine was still in perfect running order.

The Norwegians found a history-making use for an even smaller seagoing Diesel. In 1910 they bought a four-cylinder, 132-kilowatt engine from Sweden's AB Diesels Motorer and installed it in a wooden schooner built in 1892. Not, perhaps, an obvious way to go about making history. But the schooner was Fridtjof Nansen's *Fram*, specially built to withstand the pressure of polar ice. In June 1910 she left Norway under the command of the explorer Roald Amundsen, bound for Madeira and Antarctica. On December 14, 1911, Amundsen became the first human being to set foot on the South Pole, and a Diesel engine had helped to get him there. (On March 13, 1912, he telegraphed the makers from Hobart, Tasmania, "Dieselmotor excellent," and that engine's descendants have been named "Polar" ever since.)

Rudolf Diesel was doubtless pleased to be invited to represent German technology at the 1911 World Congress of Mechanical Engineers in London. He was co-guest of honor with Sir Charles Parsons, the inventor of the compound steam turbine. The two men had a deep appreciation of each other's work and soon became social as well as professional friends. Parsons was notorious for publicizing his invention with a large-scale prank: He brought his triple-turbine powered speedboat *Turbinia* to the 1897 Spithead Naval Review, an event of extreme pomp and solemnity, without an invitation. *Turbinia* raced up and down the lines of battleships at the unheard of (and uncatchable) speed of 64 kilometers per hour, making a laughing stock of the naval vessels sent after her. Following the scandal, the hitherto chilly British Admiralty conceded that the compound turbine might have some use after all.

By 1911 the Admiralty, along with the other navies of Europe, had begun to reach the same conclusion about Diesel's invention. Diesel concentrated on ships like *Selandia*. The new Danish ship behaved beautifully on her trials, and on February 22, 1912, she began a 35,000-kilometer maiden voyage to Bangkok. *Selandia* measured 113 meters long by 16 meters beam, and her displacement of just under 10,000 tonnes made her by a wide margin the largest Diesel-powered object to date. B & W fitted her with two reversible four-stroke engines conservatively rated at 932 kilowatts each at 140 r.p.m. Her sea speed was 21 km/hr—fast for the day—and she carried enough fuel oil in her double bottom to drive her 55,000 kilometers at that speed.

Selandia's external design matched the novelty of her engine room. Instead of the usual coal-black, her hull was painted a pale buff with green boot-topping, and her handsome deckhouses were arranged in a unique four-island configuration. The most startling thing about her appearance was the absence of a funnel, unnecessary since the nearly invisible Diesel exhaust was discharged from vents in the hollow mizzenmast. Sailors used to identifying steamships by their tall stacks were nonplused by *Selandia*'s streamlined profile.

Her first voyage was a triumph. She fully vindicated the

confidence of her owners, her builders, and the inventor of her engines, carrying cargo faster, farther, and cleaner than steam-powered freighters, on less fuel and without any stops for bunkering. When she called in London she was visited by Winston Churchill, then First Lord of the Admiralty, who commented that the Danish ship represented "... an advance which will be epochal in the development of shipping. This new type of ship is the most perfect maritime masterpiece of the century." H. N. Andersen, founding director of the Danish East Asiatic Company, agreed with Churchill, and before *Selandia* returned home she had two Diesel-powered sister ships, *Fionia* and *Jutlandia*. *Selandia*'s performance on her maiden voyage was not a fluke. In her first twelve years of service she sailed more than a million kilometers, and during the entire time she required only ten days for repairs. The East Asiatic Company never ordered another steamship.

About a month after *Selandia* started for Bangkok, Rudolf and Martha Diesel sailed for New York. Their visit was suggested by Adolphus Busch, who once again wanted Diesel's help with the floundering American Diesel Engine Company. The consulting trip lengthened into a two-month tour punctuated by speeches and interviews. Diesel was courteous to reporters but fatigued by their persistent questions. During a two-week stay in St. Louis he used an address at Mechanics Hall to sum up the progress of his engine so far. He now felt that the United States offered fertile ground for his invention, even though the newspapers reported (incorrectly) that there were only ten Diesel engines in the whole country. But in the world, Diesel pointed out, there were already 70,000, producing more than a million kilowatts.

He used the example of ship propulsion to show the practicality of his rational heat engine. *Selandia*'s performance wasn't known yet, but Diesel cited the conversion of the icebreaker *North Pole* to Diesel power. The new installation had reduced fuel storage requirements by 85 percent, reduced fuel weight by 80 percent, and increased the cruising range by 5,000 kilometers. The icebreaker's improved performance was typical of the several hundred vessels known to be equipped with Diesel engines at that point, including 60 freighters of various sizes.

Another area that Diesel dwelt on at length was still a black sheep. Sulzers, Krupp, and the Prussian & Saxon State Railways had spent more than two million marks over the preceding five years to develop a prototype Diesel locomotive. Their 80-tonne direct drive test machine was powered by a V-4 two-cycle engine, and was plagued with ignition and transmission problems; a flop. Nevertheless, Diesel predicted that the "thermo-locomotive" would eventually revolutionize railroading throughout the world.

He made three other observations during his American visit that showed great foresight. First, he insisted that air pollution was pernicious and would become an important consideration in future engine designs. Second, he said that despite its slow start, the Diesel locomotive would be utilized more in the United States than anywhere else. Third, he cautioned Americans that they must learn to think about efficient operation: "The leading idea in Europe is low operating cost; the leading idea in the American economy is *first* cost. The word 'efficiency,' which is the basis of every contract in Europe, is unknown to a vast proportion of Americans . . . to businessmen, and to purchasers of engines." He felt that America was "monstrously" rich in natural resources but predicted that eventually economical use of fuel must transcend the whim or wasteful habits of any country, even this one.

These pronouncements were neither fashionable nor widespread, and they were made at a time when few Americans were willing to listen to them. Every one of them has proved to be correct. It is a dogma of fortune tellers that one pays for the power to predict the future by being forbidden to profit from it. That describes Rudolf Diesel's condition very well. When he and Martha returned to München at the end of May, they were greeted by a happy daughter and son-in-law, and their first grandchild. But Rudolf's debts were worse than ever, and a prominent real estate company brought a lawsuit against him for payment on some of his speculations. He lost the suit; the payment was 600,000 marks in cash.

Fifteen

The lawsuit was almost the last straw. By June of 1912 Diesel's cumulative losses amounted to nearly 10 million marks. His book on the origins of the Diesel engine was published early in 1913, but it made only a small contribution to his dwindling credit. And not even the most blinkered idealist could ignore the political tensions of Europe any longer. As 1912 ended the Balkans were seething with declared and undeclared wars, and the industrialized nations accelerated their arms race. It was a difficult time for a pacifist. Statistics showed more than 1,500,000 Diesel kilowatts in use worldwide, but Rudolf's headaches and gout worsened again, and flares of temper punctuated his depressions.

It was not all deserved punishment. The Diesels' son Rudolf had no interest that overlapped his father's, and his temperament was morose. He left school at nineteen and moved away from the family, determined to earn his living as a clerk. The elder Rudolf was surely as hurt by this behavior as by his continued ill-fortune. And yet, as it became clear that things hadn't turned out as he had hoped, and that his chance to be a rich man was gone forever, he changed in an attractive way. During the winter of 1912–13 his interest in art and music revived, and he enjoyed simple outings to lakes and the mountains more than the earlier expeditions to grand hotels. After he visited the Sulzers in Winterthur, Frau Sulzer-Imhoof commented, "That wasn't the proud Diesel any more." Like others, she didn't disapprove of the change.

To help pay his creditors Diesel sold the red touring car; somehow there was still money to live on. He and Martha went to see more Wagner operas, and he began to study the works of Arthur Schopenhauer, the German philosopher whose somber ideas Wagner had woven into *Tristan und*

Isolde. Early in 1913, despite the resurgence of the Balkan War in the eastern Mediterranean, Rudolf and Martha traveled to Sicily. On the way home Rudolf made a detour to the Bavarian Alps to climb the Säuling, a mountain that he had climbed as a young man. This time he was fifty-five. His guide remembered that he spent a long time on the mountain top gazing at the snow-covered peaks around them.

By that time Diesel's house was mortgaged to defray some of the ever-present debts, but there were still the satisfactions of family. The rebellious Rudolf, Jr. had married, and his wife was expecting a child. The Diesels' younger son, Eugen, was growing up with an intense desire to follow in his father's footsteps. He planned to spend the next year as an apprentice at the Sulzer plant—another Diesel *blaue Monteur* —and Rudolf was able to give him a scraped-together check for ten thousand marks as a study fund. And there was always Hedy, the perfect daughter, married to the promising Arnold von Schmidt with what appeared to be both love and security.

In late spring Rudolf went to visit his sister Emma, now living in Rogätz, and on his way home stopped at the Leipzig International Architecture Exposition. Up to that point in his life he had steadfastly refused to set foot in the Graf von Zeppelin's airships. Diesel's caution was well founded because most of the Count's early machines were destroyed in accidents after a few weeks of flying. Astonishingly not one life was lost in this series of disasters, but the Zeppelin Company went through a roller-coaster history that made Diesel's fortunes look almost placid. Slowly the big rigid airships were improved, and in 1910 the first passenger airline in the world, DELAG (*Deutsche Luftschiffharts Aktien Gesellschaft*) was founded in Germany, using Zeppelins. In the spring of 1913 DELAG's newest ship, LZ17 *Sachsen*, was giving two-hour sightseeing flights at Leipzig, and Diesel agreed to go aboard her. *Sachsen*, powered by three 132-kilowatt Maybach gasoline engines, was a huge machine even by today's standards: 126 meters long and 15 meters in diameter. (For comparison, the largest Boeing 747 is 70.5 meters long and 6.5 meters in diameter.) Diesel may have been reassured because DELAG airships had carried Orville Wright and the Crown Prince of Germany, but it is more likely that his new philosophical mood was responsible for the flight. In any case, he loved it.

He returned to München full of his adventure, and no one would have judged from the social calendar at Maria-Theresia-Strasse 32 that the Diesels were suffering any privation. During June they hosted a series of dinners for visiting engineers. The guests included Colonel Meier from St. Louis, George Carels of Carels Frères in Gent—one of Rudolf's favorite guests—and Sir Charles Parsons from England. The parties were very successful; in a period accustomed to overstuffed entertainment, each of these men made special mention in their writings of the "wonderful shindigs" [Meier] at the Diesels' home that June. The friendship between Rudolf and Sir Charles Parsons had grown into a particularly warm and sympathetic relationship. Parsons noted that Diesel was worried about his finances, and was also deeply concerned about the threat of a European war.

It was like another world: Parsons was the inventor of the most efficient external combustion engine ever known; Diesel was the inventor of the most efficient internal combustion engine. Their friendship spanned the barriers of nationality and language, and while they dined together, 56,000-kilowatt Parsons turbines were being built into the five British battleships of the *Queen Elizabeth* class—the fastest, most heavily armed and armored warships in the world—and pairs of ultralight 1250-kilowatt Diesel engines were being installed in the *Unterseebooten U-19* to *U-23*, the first Diesel-powered submarines of the German Navy.

While in München Parsons invited Diesel to dinner in England on the following October 1; Diesel accepted cheerfully. It looked as though it would fit in very well with his plans for the autumn. He had accepted a position as a consultant to the British Diesel Company, and they had invited him to be the guest of honor at the ground-breaking ceremonies for their new factory at Ipswich early in October. George Carels was a director of the British Diesel Company and was going to England at the same time. He arranged for Diesel to address the Royal Automobile Club of London on September 30 and invited him to visit Gent beforehand; they could go over together. It was a capital plan, and the three friends parted with much good feeling. A few days later, on June 30, the Reichstag passed the 1913 Army and Finance Bills. Because of the threatening military buildup of other powers, the German government was authorized to increase the strength of the army to 870,000 men, with corresponding increases in armaments. The cost was estimated at one billion marks.

The Balkan countries had no time to analyze military postures. While the Reichstag debated, the Second Balkan War broke out and raged through the summer, until the Treaty of Bucharest nominally ended it on August 10. It was like trying to cap a volcano with sealing wax. Rudolf saw what he dreaded happening around him and became more and more depressed. At the beginning of September Martha decided to visit her mother in Remscheid. Remscheid is in the northwest part of Germany, due east of Gent; the tentative plan was for Rudolf to meet Martha at Hedy's house in Frankfurt and take her with him to England.

As soon as Martha left, Rudolf sent Eugen off to Winterthur to begin his apprenticeship, gave the servants a long weekend holiday, and invited his eldest son to spend a few days with him at Maria-Theresia-Strasse. Rudolf, Jr. was surprised at the tenor of the visit. His father led him through the house with a ring of keys, explaining what each one opened, and what was in the various closets and drawers. He made a point of showing him where valuable papers were kept. The two men took a day trip to the picturesque Starnberger See a few kilometers south of München, but neither of them was in a lighthearted mood. When the servants returned, they noticed that many papers had been burned in one of the furnaces.

On September 12 Rudolf took a train to Frankfurt and Hedy's house. The next two weeks were like a sunny remission; he and Martha played with their two grandchildren, and Rudolf toured the Adler Works where his son-in-law was a director. He charmed von Schmidt's parents, and frightened Martha by driving an Adler-built automobile with great élan. He was very tender toward his wife during this visit, but sometime during the fortnight he decided to go to England without her. Frankfurt is famous throughout Germany for the quality of its leather goods. Rudolf, as the son of a *maroquinier*, was a knowledgeable connoisseur of them, and it seemed appropriate that his going-away present to Martha was a handsome overnight case. He was a little odd about it, insisting that she take good care of it, and that she not open it until the following week. She promised to oblige him.

On the 26th Rudolf traveled to Belgium by slow train, the sole occupant of a first-class compartment. His reading was not light: Schopenhauer's *Parerga und Paralipomena.* In Gent he checked into the Hotel de la Poste, and on the 27th wrote his wife a loving and confused letter, mentioning that he hoped for word from her in London. He had met Martha 31 years earlier as the young German governess of a French family. He had been delighted then by both the poetic quality of her German, and her familiarity with the works of French writers like Victor Hugo. Perhaps remembering, he used both languages in the poignant last line of his letter: "*En attendant je t'aime, je t'aime, je t'aime. Dein Mann.*" The letter was misaddressed to her, care of Arnold von Schmidt, Maria-Theresia Str. 32, Frankfurt/M. There was no such address in Frankfurt, and after a series of detours the letter eventually reached her on October 5.

On the 28th Diesel wrote to his son Rudolf and noted that he had been suffering from severe headaches, insomnia, and heart pain. The next afternoon he boarded the channel steamer *Dresden* at Antwerp with George Carels and Carels' chief engineer Alfred Laukman. The three men had an excellent, leisurely dinner together. Carels later reported that Diesel seemed to be in good spirits, but he also spent much time reassuring the inventor that his debts were only temporary. After dinner they took a turn around the promenade deck, and at about ten o'clock retired to their separate staterooms. The crossing was smooth and storm-free, but when Carels and Laukman met at breakfast, Diesel did not appear. Carels went to Diesel's stateroom; the bed had not been slept in, and the luggage was untouched. When he reported the disappearance to the officer in command, he learned that a petty officer had found Diesel's hat and coat neatly folded under the stern railing. Captain H. Hubert ordered the ship held at sea for search, but no trace of the missing inventor was found. The *Dresden* docked at Harwich one hour late, and the captain delivered to the master of the port a missing-at-sea certificate for Dr. R. Diesel.

The disappearance was reported to the German vice-consul at Harwich on October 1st. There was little evidence to glean: the purser reported that Diesel had left a call for 6:15 A.M., but the inventor's notebook on his bedside table had a small cross penciled in it after the date September 29. His key ring hung from the lock of his suitcase. On October 10, the Belgian pilot steamer *Coertsen* sighted a body floating on the sea, and the crew took from it a coin purse, a medicine kit, and a spectacle case. The body was left at sea, and the articles were delivered to the master of the Dutch port of Vlissingen on the same date. Eugen Diesel identified all three of them as his father's.

When Martha Diesel opened the overnight bag Rudolf had given her, she found twenty thousand marks in cash. An audit of Diesel's finances as of October 1, 1913 showed every bank account empty, no reserve cash, and many debts for interest on loans. Martha and her sons had little difficulty understanding that Rudolf had taken his own life. As is often the case when a celebrity dies, the sorrow was tarnished by a burst of lurid speculation in the popular press. Not many of the media scandalmongers could have read the man that Diesel took as his preceptor. Schopenhauer believed that denial of the life-will, with its acquisitive taints, was the purest deed of all. Richard Wagner, Diesel's favorite composer, dramatized in Isolde's love-death what Schopenhauer taught: Suicide is the ultimate affirmative act. Diesel had worked hard, had succeeded, and yet had failed. The sad probability is that in the end he adopted Schopenhauer's ethic to save what was left of his pride and self-respect.

Sixteen

War is usually a forcing ground for technology. An invention often evolves most rapidly when it is new. Both of these principles acted at the same time for the Diesel engine. Nine months after Rudolf Diesel disappeared from the *Dresden*, the Archduke Francis Ferdinand of Austria was murdered at Sarajevo, and one month after that most of the nations of Europe were at war. It is unlikely that Diesel would have approved of the way his invention was used in World War I, or the price paid for its development.

In August 1914, when many of the belligerents took sides, the Diesel engine was still a relatively new machine. One indication of its place near the advancing edge of technology was the cost of building it; a contemporary rule of thumb priced a Diesel kilowatt about the same as three gasoline-fueled kilowatts. Part of the difference was in manufacturing cost; only skilled machinists using first-class tools could hold the tolerances necessary for the Diesel cycle. The rest was in materials, because Diesels were heavier than comparable-sized gasoline engines, and required high quality castings and steels. Their advantages were still compelling enough to encourage plenty of research.

Two areas of research that continued to yield important results were reversing engines and two-stroke design. Sulzer's early experiments with the two-stroke gave them a significant head start; by 1914 they had developed scavenging and piston cooling systems that allowed them to build two-cycle Diesels rated at 750 kilowatts. Their reversible two-cycle of 1905 also stimulated manufacturers of four-cycle engines to solve the reversing problem. The first reversible four-stroke Diesel was an 88-kilowatt paddle tug engine built by Nobel Brothers in 1907. This engine led to a contract with the Russian Admiralty, and a famous naval incident.

Nobel proposed to supply two 365-kilowatt reversing Diesels for each of the new Caspian Sea gunboats *Kars* and *Ardagan.* The gunboats were to protect Russian shipping from Persian pirates, and oil engines were specified so the ships could leave port quickly, without the usual time to raise steam. The contract was fail-safe; if the engines didn't operate as promised the Nobels were to convert the ships to electric reversing at their own expense, involving a loss of about $100,000 per ship. During the trials of the *Kars* in the Gulf of Finland the captain raised the Russian naval ensign, thus taking official responsibility for the ship.

As the *Kars* entered the River Neva, the helmsman mistakenly headed the ship straight for the granite embankment with no room to turn. The captain, used to steam machinery, lost his nerve and refused to reverse the new engines. At the last moment Dr. M. P. Seiliger, a civilian Nobel director who was observing on the bridge, threw the handles of both engine-room telegraphs to Full Astern. Before the signal was acknowledged from below, the ship slowed down, came to a shuddering halt just short of the wall and surged astern, demonstrating the advantage of quick reversing in an unequivocal way.

The lesson of the *Kars* was not lost on naval architects. By 1914 every modern navy was experimenting with Diesels. The designs ranged from lightweight aluminum engines built for British auxiliary boats, to Krupps' huge warship engine prototype, which produced 8,832 kilowatts from six cylinders in 1915. But it was in the submarine that the Diesel was to become an instrument of world strategy. During the first decade of the twentieth century, the navies of Britain, France, Italy, Japan, Russia, and the United States built gasoline-engined submersibles and found them wanting. (Germany avoided the hazards of petrol fumes in a sealed hull from the start, and powered her first submarines with heavy oil engines.)

The adoption of Diesels made immediate and dramatic

changes in the nature of the Submarine. The United States Navy H Class boats of 1911-13 had gasoline engines that gave them a range of 3,700 kilometers at 20.4 kilometers per hour. The overlapping K Class, with Diesel-powered hulls of similar size, had double the range at the same speed. Comparable advances were made in many fleets, but the cramped, oily submarine was still a dubious weapon for flag officers used to the polished brass and broad decks of a battleship.

The turning point came on September 22, 1914, when the German submarine *U-9*, under Kapitän-Leutnant Weddigen, encountered three British cruisers, *H.M.S. Aboukir*, *Cressy*, and *Hogue*, 48 kilometers off the Dutch coast. In one hour Weddigen sent all three ships to the bottom. *U-9* displaced 508 tonnes and had a crew of 29 men; the combined displacement of the three cruisers was 36,577 tonnes, and of their crews 800 men were saved and more than 1,400 drowned. Even the most reactionary admiral could not face such dismal statistics and ignore the submarine.

That meant that the governments of the warring countries could not ignore the prime mover of submarine warfare. The international cooperation that Rudolf Diesel prized so highly disappeared and was replaced by feverish and secret competition. The result was a kind of mechanical Babel in which engine designs took on national peculiarities, though they were all directed toward the same goals of higher power and lower specific weight. In many countries specific weight was driven down from the hundreds of kilograms per kilowatt of Diesel's prototypes to fifty kilograms per kilowatt and less. By the end of the war the compatibility of the Diesel engine and the submersible warship was frighteningly clear. One hundred seventy-eight German U-boats were sunk between 1914 and 1918; during the same period Britain alone lost 7,883,235 tonnes of shipping, almost all of it to submarines.

The search for lower engine weights produced some remarkable offshoots. In 1913 the brilliant aeronautical engineer Hugo Junkers built a four-cylinder Diesel aircraft engine. A six-cylinder model developing 368 kilowatts at 2,400 r.p.m. ran in 1916. The problems of metamorphosing the ponderous stationary Diesel into an airborne prime mover were of the same order of difficulty as Diesel's original ones. Nevertheless, Junkers persisted, and ten years later his company unveiled the *Jumo*, a six-cyclinder, two-cycle, opposed-piston aircraft Diesel that powered the Dornier Do18 flying boat and other aircraft. (In March 1938 a Deutsche Lufthansa Do18 broke the world long-distance flight record when it flew nonstop from the English Channel to Caravellas, Brasil—8,392 kilometers.)

Besides spurring competition between opposing design teams, the war also separated the types of engines built by belligerent and neutral countries. In 1914 Scandinavians were renowned as superb sailors but not as heavy manufacturers. They had the motorship market virtually to themselves during World War I and were able to establish a preeminent position. Despite the heavy losses of merchant vessels between 1914 and 1918, the number of Diesel-powered ships rose from 297 to 912 during the same years, and their combined tonnage more than tripled. Many of them were built in the noncombatant northern countries.

North America was even more isolated than Scandinavia from the skirmishing that preceded the war. Adolphus Busch's American Diesel Engine Company went through several reorganizations between 1902 and 1912, but sales and development moved very slowly compared with progress in Europe. In 1912 the United States Diesel patent expired, and many American Diesel engine factories were founded in the next two years: Allis-Chalmers, in Milwaukee, Wisconsin; Fairbanks Morse, in Beloit, Wisconsin; Nordberg Diesel, also in Milwaukee, and Worthington, in Cudahy, Wisconsin. Rudolf Diesel often joked that his engines could run on butter, and at first glance the clustering of so many Diesel manufacturers in America's most important dairy state seems to bear him out.

There was, of course, another reason. After the first

wave of German immigrants to Pennsylvania (the Pennsylvania "Dutch") in the late seventeenth and early eighteenth centuries, thousands of German settlers moved to the middle of the country from the 1830s on and founded farms and businesses. In 1900, 72 percent of the population of Milwaukee was of Germanic origin, and it was among this German-oriented and often German-speaking group that Diesel's invention flourished. The sites of other U.S. Diesel factories—Busch-Sulzer in St. Louis and Winton in Cleveland, Ohio, also had substantial German immigrant populations. (Since the brewing industry in America established this same trail, it has been waggishly suggested that Diesel was mistaken and his engine actually ran on beer.)

The first American Diesels were, with rare exceptions, behind contemporary European engines technologically. The supplies of oil and iron ore seemed unlimited, and skilled labor was scarce and expensive, so American factories tended to build rugged engines that were not particularly low in fuel consumption. At the end of 1913 there were about 62,000 Diesel kilowatts in use or on order in the United States. Engine production increased tremendously during World War I, but design sophistication came very slowly.

Perhaps part of this slow evolution can be traced to the same kind of chauvinism that troubled Rudolf Diesel so many times. During World War I some American superpatriots used the war as an excuse to indulge in campaigns against their German co-immigrants fighting the same war on the same side. The resulting pressures tended to reinforce the geographical isolation of the Midwest. The habits of designing heavy, thirsty engines became deeply entrenched, and a decade and more after the armistice, American journals were still hailing some local Diesel advance that had long been passed elsewhere. The differences were emphasized in the 1920s, because America and Europe were both becoming aware of a new use for the Diesel: land transportation.

Seventeen

The internal combustion engine came out of its adolescence with Daimler and Benz's prototypes of the automobile. Between 1885 and 1914 the Western world was seduced by this device as by no other mechanical invention, and its extraordinary progress is a measure of the value that people place on quick, private transport overland. By the eve of World War I, Benz's three-wheeled dogcart had evolved into the car we are familiar with today in all but detail refinements.

If that sounds immoderate, here is a short list of automobile features that were invented, patented, and road-tested before 1914: Steering wheel (Vacheron, 1894); Pneumatic tires (Michelin, 1895); Front-wheel drive (Gräf und Stift, 1897); Honeycomb radiator, gate gearshift, foot accelerator (Daimler, 1899); Shaft drive with universal joints, live, sprung rear axle (Renault, 1899); Disc brakes (Lanchester, 1902); Fuel injection (Bollée, 1902); Independent rear axle suspension (Rumpler/Adler, 1903); Overhead camshaft engine (Maudslay, 1903), V8 engine (Ader, 1903); Shock absorbers (Mors, 1903); Automatic transmission (Sturtevant, 1904); Coil and distributor ignition (Delco, 1908); Transverse engine (Christie, 1909); Safety glass (Triplex, 1909); Electric starter and electric lights (Cadillac, 1911); Four-wheel servo-assisted brakes (Isotta-Fraschini, 1911); All-steel body (Oakland and Hupmobile, 1912); Double-overhead camshaft engine (Peugeot, 1912).

The list could easily be doubled, but this should be long enough to show that the first 30 years of automobile design were the most creative. The increase in production was also impressive. The total world production of cars in 1903 was just under 62,000; in 1914 Henry Ford alone sold 248,307 vehicles. Despite this, the 1914 automobile was still associated with affluence rather than utility, and it was a male preserve.

Four years later every modern army was mechanized, and thousands of men and women who could not have afforded to own automobiles had become skilled drivers of staff cars, ambulances, and heavy trucks. Manufacturers on both sides of the conflict had been forced to standardize production as never before, and had learned to build genuinely interchangeable engine parts at strictly controlled prices. Two long-established prime movers—the horse and the steam engine—had lost the war to internal combustion.

Given this background it is hardly surprising that Rudolf Diesel wanted to adapt his invention to the motor vehicle. At the time he tried to do it, the Diesel was too heavy and too complicated to compete with the gasoline engine. The air-blast injection system with its unwieldy compressor was a major drawback, and the high torque/low speed characteristic of the Diesel didn't mesh with the evolving concept of the automobile. Karl Benz, among others, persisted where Diesel had left off, and it was in the workshops of Benz & Cie that Prosper L'Orange eventually combined many ideas into a successful "solid" fuel injection system. In this system a small high-pressure pump replaces the air compressor, moving the liquid fuel through tubes to the injectors. As each piston reaches the top of its compression stroke a precise quantity of fuel is forced through the appropriate injector, vaporizes, and ignites—in L'Orange's design in the precombustion chamber. It is considerably more difficult to accomplish this than to describe it.

L'Orange had started on the problem in 1909 and made slow progress, but between 1914 and 1918 the German government requisitioned all of his time for designing U-boat engines. Even during this time he tried to keep in touch with engineers in other countries who were working on fuel injection, and his letters give the impression that it was more a matter of personal preoccupation than national defense.

Before the war two Swedish engineers, Jonas Hesselman, chief engineer of AB Diesels Motorer in Stockholm, and Harry

Leissner at Ljussne-Woxna, also began to have some success with airless injection. Hesselman had been only one year behind Sulzers in developing a reversible two-stroke marine Diesel, and he designed a reversing four-stroke in 1909. In that year, at L'Orange's urging, Benz acquired the license for one of Hesselman's injector designs (the engine of the *Fram* had Hesselman injectors), but the Benz engineers only learned about Leissner's work after the armistice. Because of the war the duration of German patents was extended by four years, and so Benz was able to buy the rights to Leissner's improvements and incorporate them in L'Orange's patented injection system of 1919.

By that year Benz had a one-cylinder research engine running smoothly up to 800 r.p.m. over a wide load range with the Leissner/L'Orange fuel injection, and in September of 1920 they delivered the first commercial Diesel engine without a compressor. Nine months later Benz began to offer direct injection Diesels in motorized plows. They were an immediate success; and on February 4, 1924, at the Amsterdam Automobile Exhibition, Benz & Cie unveiled the first series production Diesel road vehicle in the world. It was a somewhat stately entrant into the jazz age of the automobile—a five-tonne open-bodied truck.

More than fifty years later the Benz 5K3 still looks as though it means business. It was built at Benz's Gaggenau division, and had an upright enclosed cab and a long, low-sided body supported on a four-tired rear axle. This patriarch of today's heavy Diesel trucks weighed 4400 kilograms loaded, and while it had no bumpers, few modern vehicles would argue with its massive frame rails and monumental radiator. The four-cylinder engine produced 36.8 kilowatts at 1,000 r.p.m., and it could move a five-tonne load at 22 kilometers per hour. These specifications were not nearly as interesting to German truck buyers as the fuel consumption. Benz had spent six months test-driving the truck and comparing it with gasoline-engined competitors of similar size and weight. A typical 100-kilometer run produced the following results: Over the same distance, carrying the same load at the same speed, the Diesel engine used 26 percent less fuel by weight, which represented a startling 86 percent saving in fuel cost. (Since fuel is usually bought by volume, not by weight, it is important to note that Diesel fuel is denser than gasoline. This gives it even greater relative economy; in the test described above the fuel saving by volume at the pump was 32 percent.)

Fuel economy was an essential virtue in post-World War I Germany, but the Diesel truck was introduced at a time when it took tremendous confidence to believe in any business venture. At the end of 1923 the value of the mark had fallen to less than the cost of the paper it was printed on. (The Rentenmark, introduced as a desperation measure on November 15, 1923, was exchangeable for one trillion of the old marks.) The government was in chaos, with ministries and cabinets changing every few months, and the collapse of the mark also caused a deflation of the French franc. Only after the Dawes Plan made it possible to reorganize the German State Bank and stabilize a new Reichsmark in August of 1924, was there a sound financial base in Western Europe.

The tone of the entries in engineering notebooks during this period indicates that it was probably helpful to have something like an engine to concentrate on. A few months after the Benz 5K3's introduction, M.A.N. put their first Diesel truck on the market, also powered by an airless-injection engine. Gottlieb Daimler and Wilhelm Maybach's pedigree in commercial vehicles was as honorable as M.A.N.'s with the Diesel; they had collaborated on the first gasoline-engined truck in the world in 1896. (Since 1900, when Maybach designed the car that most experts regard as the forerunner of the modern automobile, Daimler's products had carried the name Mercedes after the daughter of Emile Jellinek, the Daimler distributor in Nice.) In 1923 Daimler's Berlin subsidiary showed a Diesel-powered bus, truck, and dumper at the Berlin Motor Exhibition, but their engines still used an air compressor. Daimler was impressed with the higher output of Benz's compressorless Diesel, and there were many other points of convergence between the two companies. In 1926 Daimler and Benz merged into Daimler-Benz AG, and began marketing their products under the name Mercedes-Benz. It was Benz's Diesel engine that went into the new company's trucks.

The Benz Diesel was equipped with a glow plug designed by Robert Bosch; it contained a filament which heated up the precombustion chamber when an electric current was passed through it. Bosch was also responsible for a crucial improvement in Diesel fuel injection. In 1925, building on the experience of the Swiss engineer Fritz Lang, he designed a simplified and improved fuel injection pump which made its debut in the Mercedes-Benz OM5 Diesel of 1927. This six-cylinder engine developed 55 kilowatts and doesn't look very different from its successors of the next five decades. Various versions of the Bosch glow plug and fuel injection system also became widely accepted components of many Diesel engines down to the present day.

As a prime mover for commercial vehicles, the Diesel was quickly successful in Europe. By 1930 M.A.N.'s entire line of trucks was Diesel-powered, and four years later $\frac{9}{10}$ of all light trucks registered in Germany were equipped with Diesel engines. Heavier Diesel installations increased to 92 percent of all trucks over five tonnes. The pioneers soon had competition from companies like Büssing, Henschel, and Krupp for the German market. Abroad, licensees proliferated; a short list of Diesel truck manufacturers before 1935 would include Austria's Gräf & Stift; Atkinson and Foden in Britain; Tatra in Czechoslovakia; the French builders Berliet, Panhard, and Renault; Alfa-Romeo, Fiat, and Lancia in Italy; Isuzu and Mitsubishi in Japan; Scania of Sweden, and Switzerland's F.B.W. and Saurer.

The obvious absentee from this list is the United States. In 1927 Mack Trucks, Inc. of Allentown, Pennsylvania introduced a precombustion chamber Diesel licensed from Lanova of München. Despite the impressive record of Diesel trucks elsewhere, and publicity from several Diesel record runs by the struggling Cummins Engine Company of Columbus, Indiana, Mack had the field to itself for nearly a decade. In 1935 Oshkosh Trucks of Oshkosh, Wisconsin, and the Sterling Motor Truck Company of West Allis, Wisconsin, offered the option of a six-cylinder Cummins Diesel in some of their models. As of that year, there were only 1,400 Diesel-powered trucks in the entire United States.

When Daimler AG displayed their omnibus in 1923, some potential customers worried that the exhaust odor of the Diesel would prevent its success as a bus engine. These fears were quickly dispelled when buses placed in service during 1931 in cities including Belfast, Budapest, and London proved both popular and economical. (The ten Diesel buses tested by the London Passenger Transport Board averaged almost double the tonne-kilometers per liter of fuel recorded by otherwise identical gasoline-engined vehicles.) By 1935 London Transport had 800 Diesel buses in service and was converting its remaining carriers at the rate of six per week. An economy of this magnitude seldom goes unrecognized for long; in August 1935 the British government raised the tax on Diesel fuel by 700 percent. The Transport Board found that it still paid to run Diesels.

The history of the Diesel bus in the United States parallels that of trucks. In 1929 a European Diesel engine was tested in an old bus chassis with electric drive by the Public Service Corporation of New Jersey. Despite a favorable report on its performance and made-to-order publicity, no Diesels were operated until well into the next decade. In the fall of 1932 C. L. Cummins had a Cummins Diesel engine installed in a ten-tonne Mack bus, and on November 13 he and four companions started across the United States from New York. They arrived in Los Angeles on the seventeenth, having driven 5,181 kilometers in 78 hours and 10 minutes.

Their average speed was faster than the schedules of transcontinental express trains, and the total fuel cost for the trip was $21.90. This demonstration made no impression at all on the American public transport industry. Some 250 Diesel-electric buses were first placed in service around the U.S. in late 1937 and 1938, a year in which there were 26,000 Diesel buses registered in England alone.

Eighteen

There were several ironies in the success of the mobile Diesel engine in the United States. First, America has the self-image of a "fast" country. When C. L. Cummins wanted to dramatize the possibilities of the automotive Diesel in 1931, he installed an industrial Diesel engine in a race car, drove it 162.1 kilometers per hour at Daytona Beach, and then entered it in the annual Indianapolis Speedway race. (The car finished in twelfth place at an average speed of 138.65 km/hr. without making a single pit call—the first car in history to run the 805 kilometers nonstop.)

Second, even though sales of the fuel-saving Diesel had burgeoned during Europe's long and crippling depression, many analysts have cited the depression economy and the shortage of capital during the 1930s as the reasons for the Diesel's slow acceptance in America. Walker Evans and Dorothea Lang's haunting photographs, and John Steinbeck's *The Grapes of Wrath* have imprinted the scourged farmer on the American conscience as a symbol of that depression. Nevertheless, the first mobile Diesel that sold in numbers in the United States was a tortoise, not a hare, and it sold during the depression to builders and farmers.

The Caterpillar 1C1 Diesel-engined tractor was introduced in 1931, and it was an instant success. The Caterpillar company, formed in 1925 by the merger of two California tractor and harvester manufacturers, hedged its bets by offering gasoline-engined and Diesel-engined models of the same nominal sizes, but 9/10 of the machines sold were Diesels even though they cost about 30 percent more. Caterpillar sold 10,000 of these crawler tractors during the next four years, and was quickly followed into the Diesel market by the Cleveland Tractor Company and International Harvester. In 1937 about three-quarters of a million Diesel kilowatts were produced in North America for tractor installations.

By that year the Diesel was also beginning to make time on rails. The Klose-Sulzer experimental locomotive of 1912 was upstaged one year later by a deft little Diesel-electric railcar built by ASEA for Sweden's Mellersta & Södermans-lands Railway. Its AB Diesels Motorer engine was built to the same standards of longevity as a marine powerplant, and this first revenue-earning Diesel railway vehicle paid its way until 1939.

The coupling of an internal combustion engine, a generator, and electric traction motors was a system that some American railroads understood very well. W. R. McKeen Jr., Superintendent of Motive Power for the Union Pacific Railroad, had been building gasoline-powered railcars since 1905 with varying degrees of success. (During his 1912 United States visit, Rudolf Diesel made a special trip to Arkansas to inspect a McKeen car in operation.) McKeen's ideas were refined by other builders, until in 1924 the newly formed Electro-Motive Company brought out a breakthrough in gasoline-electric drive. Their 59-passenger coach was powered by a Winton engine, and it could work a branch passenger line for about half the cost of steam locomotives, day in, day out.

During this same period the Diesel engine was beginning a slow penetration of the American locomotive market. In 1923 General Electric collaborated with American Locomotive Company and Ingersoll-Rand on an experimental Diesel switcher tested in the New York Central's yards. This unit could work around the clock for only 10 cents per kilometer; it did the same work on one tank car of fuel oil as a comparable steam switcher did on twelve cars of coal, used no water, and produced neither smoke nor ashes. A convincing performance, one would have thought, but it showed instead that Rudolf Diesel had been right when he observed the American preoccupation with first cost instead of efficiency. The Diesel demonstrator was priced at $100,000, double the cost of a large steam switcher, and no one bought it. Two years later the same consortium built four improved versions of their demonstrator, and this time the Jersey Central Railroad bought the first commercially produced American Diesel locomotive and numbered it 1000. It served at their Bronx Terminal for 32 years.

Despite such performances, the Diesel locomotive was not able to overcome objections to its cost for a decade. Milestones like the Canadian National's 1925 Diesel-electric railcar fleet, or the first-ever Diesel road locomotive designed for the German State Railways in the same year were noticed mostly by railwaymen and small boys, but in May, 1932, a Diesel train became a subject for tabloids and newsreels. This was the German Reichsbahn's *Flying Hamburger*, a two-unit Diesel-electric streamliner. The body shape was determined by wind tunnel tests made at the Zeppelin Aircraft Works, and the power units were two Maybach V-12 engines developing 300 kilowatts each. On its trials the train reached 198.5 kilometers per hour, and it ran 160 k.p.h. on a daily basis during its first year of service.

Imitation is the sincerest form of flattery. Two years later the original train was still booked solid, and the German Railways had thirteen Diesel streamliners on order. Similar new lightweight expresses were being built or put into service in Belgium, Britain, Czechoslovakia, Denmark, France, Holland, Hungary, Italy, Rumania, and on four American railroads. These were the Union Pacific ("City of Salina"), the Chicago, Burlington & Quincy ("Pioneer Zephyr"), the New York, New Haven & Hartford ("Comet"), and the Gulf, Mobile & Northern ("Rebel"). Most of these trains were handsome, and many of them incorporated air-smoothing, modern materials, and tasteful interiors. To the public, with the help of some media image-making, they were "The trains of the future." The streamliners consolidated that image with performances like the Pioneer Zephyr's nonstop run from Denver, Colorado, to Chicago, Illinois, in thirteen hours on May 26, 1934. The all-stainless steel train averaged 125 kilometers per hour, and finished its 1,633-kilometer trip at the Century of Progress Exposition, having broken every railway speed and distance record in existence.

This time it was the laymen who believed, and the train men who shook their heads. The streamliners weren't real trains, for all that. They were lightweight, slick, specially maintained. To move real loads—freight—you needed an engine with muscle, and to most railwaymen all over the world that meant only one thing: steam.

This idea was understandably regarded as reactionary by the companies that built Diesels. In 1930 both the Electro-Motive Company and the Winton Engine Company were bought by General Motors Corporation. The resulting Electro-Motive Division began an aggressive campaign to improve the performance and image of the Diesel as a railway prime mover. Their first success was the lightweight engine that powered the Pioneer Zephyr on its record run. Their next target was much more ambitious: slaying the fiery dragon of railroading, the steam locomotive.

General Motors No. 103 was specifically built for that task. It was designed to compete with the largest and most powerful steam locomotives in existence, but it made no concessions at all to their brute-force esthetics. Its four articulated units blended smoothly into a 59-meter-long machine that looked suspiciously like one of the lightweight streamliners. The paint sold to modelers of American steam locomotives is called with more truth than exaggeration "grimy black"; the 103 was painted a polished dark green with two brilliant yellow bands running its entire length. At each end was a high cab set over a bow that gave an impression of grace and purposefulness combined, and the yellow sidebands swept down over the prow to form a V framing the stylized letters "GM."

These refinements caught the imagination of the public, but like the immaculate, temperature-controlled cab with its flight-age controls, they only increased the apprehension and skepticism of old-line railroad men. Fortunately EMD had set out to make a legend, not a movie, and 103's machinery was even more impressive than its styling.

Each of its four units contained a sixteen cylinder V-type Diesel engine rated at just over 1,000 kilowatts, and an EMD generator; the generator powered the traction motors on each axle of the two four-wheel trucks. The end cab units were designated "A," the middle units "B." In GM's nomenclature the 103 was thus an A-B-B-A locomotive with a nominal power rating of 4,000 kilowatts. It weighed 408,000 kilograms, making it the heaviest locomotive in history as well as the longest, but its weight was distributed over 16 axles, all driving. The four units together carried 18,000 liters of Diesel oil, enough for 805 kilometers running between refuelings. The 103 had one ability that no steam locomotive could duplicate: In a few moments it could divide itself, amoebalike, into two independent engines of 2,000 kilowatts each.

Although EMD's demonstrator was assembled from tested components, that wasn't a guarantee of success. One indication of the new engine's capabilities came on the first dynamometer runs, when it developed a tractive effort of 100,000 kilograms, twice that of the largest American steam locomotives. On a cold, blustery November 25, 1939, No. 103 left La Grange, Illinois, on its mission. In the next year the engine ran 134,805 kilometers on twenty Class One railroads in 35 states, at altitudes from sea level to 3100 meters and temperatures from −40°C to +43.3°C. Not once during this time was a single delay charged to the locomotive (that is, it had 100 percent availability), and in every test it outperformed steam engines dramatically.

Not only did it pull longer, heavier trains at higher speeds, it proved far more efficient on slippery track, grades, curves, and snow-covered right-of-way. Experienced engineers had to be cautioned repeatedly by EMD's instructors not to back the engine on a grade or take out coupler slack to get a running start as they were used to doing with steam engines. The 103's starting torque was so great that it would break the train apart if this was done. (It took a string of broken trains across the United States to make that point.) During one trial in New Mexico a conductor at the rear of a freight train hauled by No. 103 applied the air brakes in the belief that the engine was slowing for a stop; at the same moment the engineer applied full throttle. The train slowed almost to a stall—the point where a steam locomotive would have begun to spin its drivers—but 103 simply kept pulling. The dynamometer in the test car behind the engine registered its absolute limit of 122,500 kilograms, the pen shot off the scale, the dynamometer exploded, and the engine ripped the steel drawbar between itself and the rest of the train in half and leaped ahead. The advocates of muscle had not much to say after that.

Railroad executives, still harassed by the depression economy, saw more than brute power in 103's performance. Right-of-way superintendents around the country reported that the Diesel's smooth application of torque was far easier on track and switches than the hammering of steam engines. And, even more significant, the Diesel's fuel cost per tonne-kilometer averaged over the whole year was exactly 49 percent of the competing oil-fired steam locomotives. No. 103 came home in triumph like a landbound *Selandia*, and when her engines were shut off at La Grange, thirteen major railroads had ordered mainline Diesel-electric locomotives of the same class.

General Motors refers to No. 103 as "The Diesel That Did It," and railway historians agree. During World War II the flexibility and high availability of Diesel locomotives in the United States underlined their advantages even more. The orders for steam locomotives by all United States railroads after wartime purchasing restrictions were lifted makes enlightening reading: 115 in 1945; 86 in 1948; 12 in 1950. By that time the only question was when one could get delivery of Diesel power and arrange the necessary financing. By the mid-1960s steam locomotives had almost disappeared from the trackage of the world's railways, and been replaced by Diesels.

The Cummins Diesel-engined race car that finished 12th
at Indianapolis in 1931—the first car in history
to run the race without making a single pit stop.
Clessie Cummins beams from the right-hand seat.

The statistics on the side of this truck were recorded
in 1932 and validated by official observers of the
American Automobile Association. They made almost
no impression on the American motor industry.

Diesel Grand Luxe—a 1925 Bentley Saloon converted to Diesel power in 1932 by Hugh and Joseph Gardner of L. Gardner & Sons Ltd. The Gardner 4LW engine gave the car a top speed of 128 kilometers per hour, and a distinguished racing team drove it to 5th place in the 1933 Monte Carlo Rally.

The Mercedes 260D, the first successful Diesel-powered production automobile in the world. Introduced at the 1936 Berlin Auto Show, it had a 4-cylinder 2.6-liter engine. Thousands were sold before production ended in 1940, and the car established an enviable reputation for economy and longevity.

A Dornier Do 18 powered by two Junkers Jumo opposed-piston Diesel engines. Lufthansa operated these graceful flying boats through the 1930s, and in March 1938 *D-ANHR* flew 8,392 kilometres nonstop from the English Channel to Caravellas, Brazil, a new world long-distance flight record.

The Junkers Jumo 205C Diesel engine as installed in another German airliner of the 1930's, the 10-passenger Junkers Ju 86.

A cutaway display version of the Jumo 205C, showing the complexity and technical virtuosity of this ultra-light two-cycle Diesel aircraft engine.

November 25, 1939. General Motors No. 103, "The Diesel That Did It," leaves La Grange, Illinois on the year-long, 135,000-kilometer odyssey that would prove its superiority over the steam locomotive. More than any other single event, it was this extended demonstration that resulted in American and eventually world railways converting from steam to Diesel power.

A section through the Mercedes 300D automobile engine shows the typical Daimler-Benz precombustion chamber, the Bosch fuel injection and glow plug, and the overhead cam-driven valves.

After World War II Daimler-Benz brought out the 170D, a Diesel car which became even more successful than the 260D. This 1953, 170SD sedan was bought used for $735, and has since been driven 1,270,000 kilometers (789,000 miles), the equivalent of 31 times around the earth. It is still in daily use and averaging 15.52 kilometers per liter of fuel (36.5 miles per gallon).

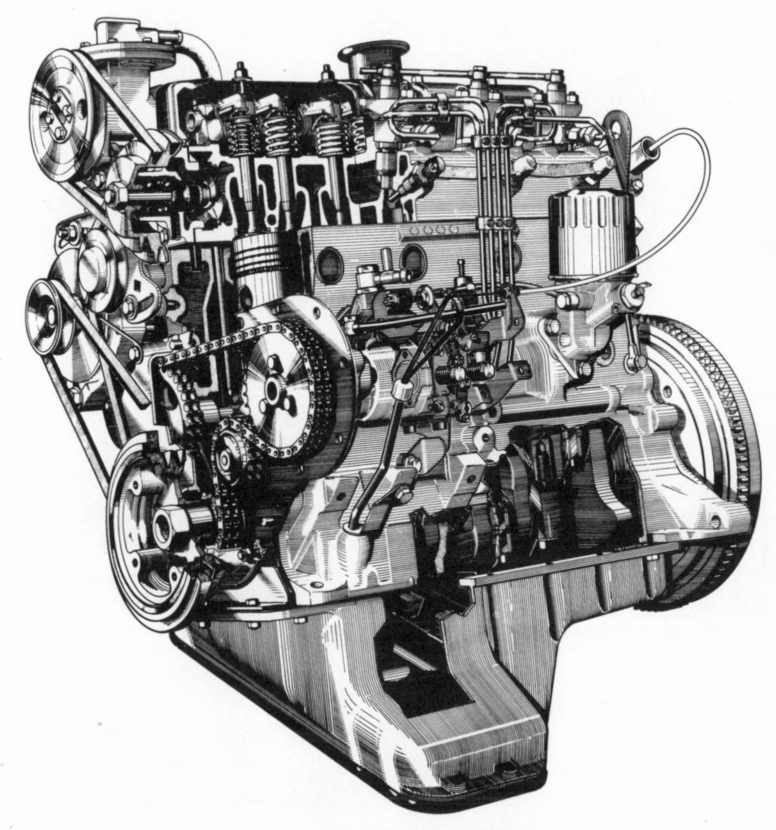

The second manufacturer to market a series-production Diesel automobile was Peugeot, who brought out the 403D sedan in 1959. The 2112cc swirl chamber engine shown in this cutaway drawing powers the U.S. version of the Peugeot 504D and develops 48 kilowatts at 4500 r.p.m.

The outside of the Volkswagen Golf/Rabbit Diesel engine . . .

. . . And the inside.

The same comparison for the Oldsmobile Diesel reveals as much about national characteristics as about engine design. At first glance this big pushrod V-8 looks little different from its gasoline-fueled American contemporaries.

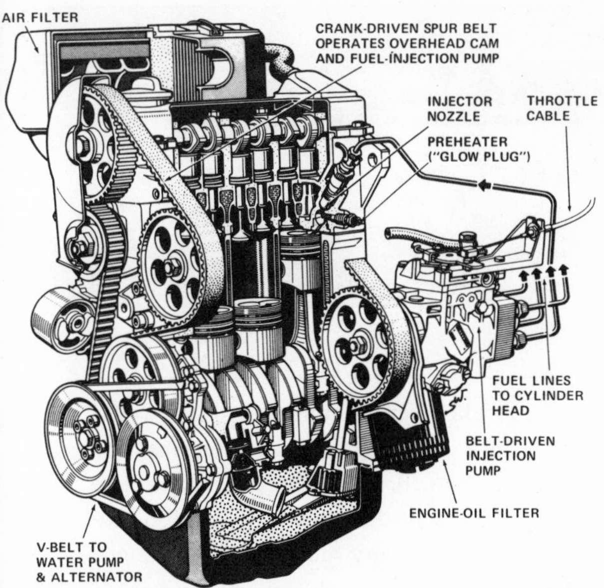

This 1471cc unit embodies many technical innovations, and produces 38 kilowatts at 5000 r.p.m., a relatively high speed for a Diesel.

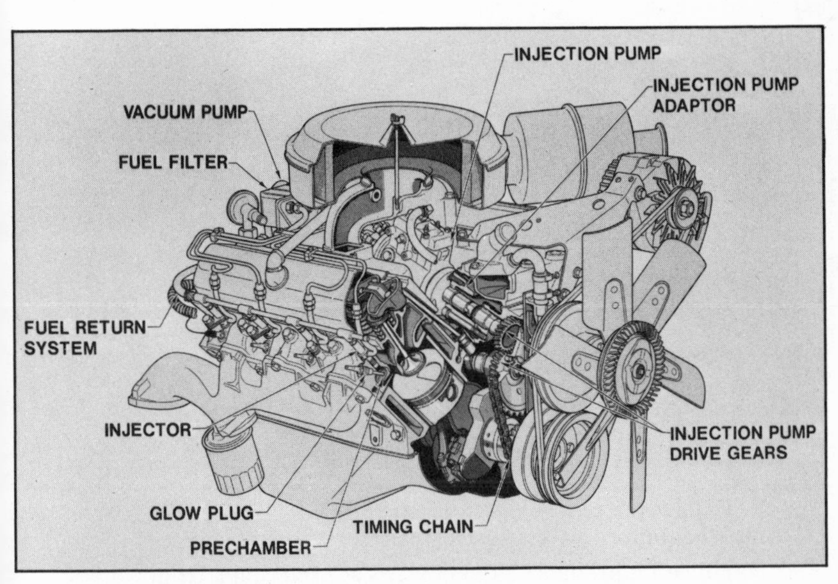

The Oldsmobile Diesel has a displacement of 5733 cubic centimeters, nearly twice that of its largest European competitors. It uses a swirl chamber designed by General Motors for quiet combustion and has an output of 89.5 kilowatts at 3600 r.p.m.

This special Opel GT broke twenty international and world speed records in June 1972. Its 2-litre Diesel engine was boosted to 70 kilowatts output by a small turbocharger, and the same engine (unsupercharged) was offered in seven Opel Rekord models in 1972, making the company the third to introduce a line of Diesel-powered automobiles.

The engine room of the Mercedes C-111: a 300D passenger car Diesel equipped with a Garrett Airesearch turbocharger. The combination produces 147 kilowatts at 4200 r.p.m. from 3 litres of displacement.

Daimler-Benz's 1976 entry into the record-breaking field. This Diesel-engined Mercedes C-111 covered more than 16,000 kilometers on the Nardo race circuit in southern Italy at 251 kilometers per hour, and averaged a scorching 253 k.p.h. for 24 hours. The three world records and thirteen international records it broke were previously held by gasoline-engined cars.

A Garrett-Airesearch TAO301 Turbocharger. This model, used for small automotive Diesels, measures about 19 centimeters long. Exhaust gas from the engine enters through the black peripheral duct, middle right, spinning the turbine wheel, and exits from the center hole to the right. Fresh air drawn into the central duct at the left is compressed by the centrifugal blower wheel and is piped to the engine through the white peripheral duct.

The Comprex Press
Wave Machine, whic
promises a major brea
through in economica
supercharging of Die
engines.

A section through a Roots positive displacement air blower, used to provide scavenging and slight supercharging in two-stroke Diesels.

A schematic diagram of t
Comprex PWS as a super
charger. The dynamic
gas pressure exchange tha
occurs in the simple-
looking rotor is actually
very complicated.

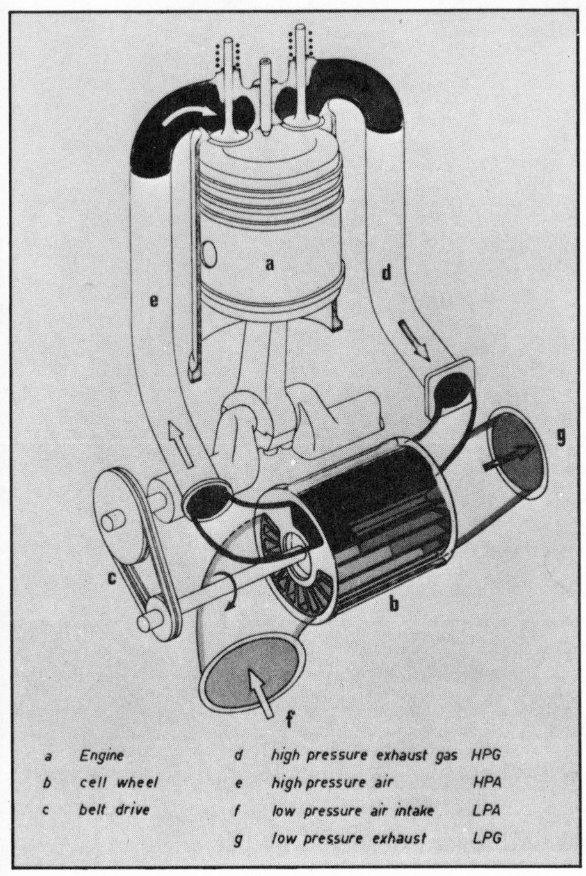

a	Engine	d	high pressure exhaust gas	HPG
b	cell wheel	e	high pressure air	HPA
c	belt drive	f	low pressure air intake	LPA
		g	low pressure exhaust	LPG

The Land of Brobdingnag. Bedding the crankshaft of a large marine Diesel at Sulzer Brothers in Winterthur, Switzerland. A single engine of this size can have an output of 35,000 kilowatts.

A similar engine completely erected and on test at M.A.N. in Augsburg, Germany. This particular two-stroke model develops 23,600 kilowatts at 106 revolutions per minute.

The current *Selandia*, descendant of Burmeister & Wain's pioneer motorship of 1912. She is five times heavier and two-and-a-half times faster than her famous namesake, and her three engines together develop 55,200 kilowatts—the largest marine Diesel powerplant in the world. As in 1912, both the ship and her engines were built by B & W.

The largest Diesel engine currently manufactured in the United States. The Delaval RV-20-4 built in Oakland, California, is a four-stroke turbocharged V-20 that develops 10,100 kilowatts at 450 r.p.m.

The Titan Diesel-electric hauler, built by the Terex Division of General Motors, weighs 553,573 kilograms loaded—one and a half times the maximum takeoff weight of a Boeing 747—and stands nearly six stories high with its body tipped up. It is powered by an Electro-Motive V-16 Diesel engine that supplies 2238 kilowatts to the truck's generator and traction motors.

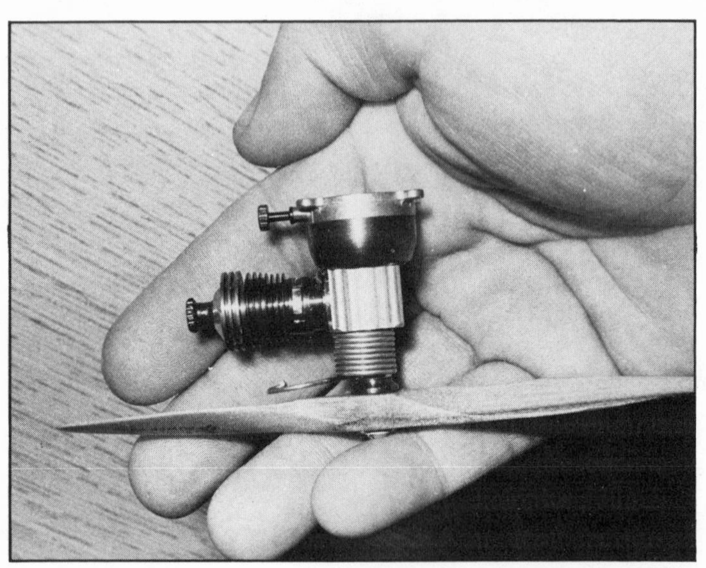

The other end of the spectrum. This Diesel engine, a conversion made by Davis Diesel Development of Milford, Connecticut, has a cylinder the size of a thimble and a displacement of 0.00032 liters!

Nineteen

If you ask an American to name the ordinary means of land transport, he will usually answer "The car," which in the United States signifies a four-wheeled, gasoline-engined, privately-owned automobile. The news that this is regarded as a distorted minority viewpoint by much of the rest of the world is likely to provoke genuine surprise, followed by defensive indignation. In fact few Americans realize that the United States, with only about 5 percent of the world's population, operates nearly 50 percent of the world's motor vehicles. Most American cars are fueled with gasoline, and many of them were designed with small regard for efficiency of any kind except profit. It has taken a somber geologic forecast and a political stranglehold on oil to make Americans begin to think seriously about the issues of waste and efficiency that Rudolf Diesel raised in 1912.

Fifteen years after Diesel's abortive trials with the automobile, Automobiles Peugeot S.A. made the same experiment. Peugeot, founded in 1890, built their first Diesel engine in 1923; and being one of the oldest manufacturers of motor cars, they naturally installed it in a car chassis. Their tests were no more encouraging than Diesel's had been. Although the company's Lille subsidiary began series production of Diesels in 1928, it built only truck and industrial engines until 1958.

The first widely publicized conversion of an automobile to Diesel power was made in Columbus, Indiana, late in 1929. It was a classically American venture in its spontaneity, its mechanical improvisation, its publicity-mindedness, and its origin in the midwestern heartland of engine manufacturing. It was done not to demonstrate engine efficiency, but to help pull an ailing manufacturer out of the red.

The manufacturer was the Cummins Diesel Company, and the man responsible for the installation was Clessie L.

Cummins, president of the company. Under his direction, a 6.25-liter Cummins industrial Diesel was shoehorned into the motor compartment of a 1925 Packard seven-passenger limousine. There was no room behind the car's radiator for a cooling fan; and even after it was removed, the clearance at the front of the engine was only a few millimeters. Between January 4 and January 6, 1930, Cummins drove this car from Indianapolis, Indiana, to New York City, arriving just at the opening of the National Automobile Show. The resulting blaze of publicity made much of the fact that the total fuel cost for the trip was $1.38.

The official reaction of the automobile industry was a cold shoulder, but Cummins was invited to demonstrate his car to the Society of Automotive Engineers in Detroit a few weeks later. Henry and Edsel Ford of Ford Motor Company, and Charles F. Kettering of General Motors also asked for, and received, private demonstrations of the Cummins hybrid. Despite these overtures and Cummins' later efforts to publicize automotive Diesel engines, not one American manufacturer put a Diesel-powered automobile into production during the next forty-five years.

Initially the situation seemed quite similar in Europe. In 1932 Hugh and Joseph Gardner of L. Gardner & Sons, Ltd. in Manchester, England duplicated Cummins' experiment by installing a Gardner 4LW Diesel in a 1925 Bentley Saloon. The Gardner Bentley could top 128 kilometers per hour (Cummins' Packard had been governor-limited to 88.5 kph), and again showed the startling economy of a well-designed Diesel engine: 273 kilometers on two shillings worth of fuel. The car was entered in the 1933 Monte Carlo Rally with a team of three drivers headed by Lord de Clifford, and completed the course fifth overall. This performance encouraged Gardners to build an experimental 3.8-liter automotive Diesel with all light alloy castings, including a magnesium sump, but they found that Britain's car builders were no more willing than America's to risk introducing a new prime mover.

Daimler-Benz, unlike Cummins and Gardner, built both cars and engines, and they did take the risk. Their first experience with a Diesel-powered car was much like their predecessors. The earliest Mercedes automotive Diesel, built in 1933, was derived from the six-cylinder truck engine of 1927. Its displacement was 3.8 liters, like Gardners' lightweight prototype, and it had a nominal output of 60.35 kilowatts at 2800 r.p.m. The plan was to install this engine in the Mercedes "Mannheim" model and introduce a Diesel-powered car at the 1934 Berlin Automobile Show. As it happened, the urge to build big proved premature; even the rigid Mannheim chassis was unable to damp out the violent detonations of the 3.8-liter Diesel, and the debut had to be cancelled.

Two years later Daimler-Benz showed that they had taken their expensive lesson seriously. At the Berlin show of 1936 they introduced a Diesel automobile with much more democratic statistics, the Mercedes 260D. This engine had four cylinders, a displacement of 2.6 liters, and an output of 33 kilowatts at 3,000 r.p.m. It was installed in the 230 chassis and fitted with a charming landaulet body—a small four-door sedan, the rear third of whose roof could be opened like a convertible.

The 260D provided exactly the combination of qualities that many European drivers were looking for. It was neither light nor fast (top speed was 97 kilometers per hour), but it would run 10.5 kilometers on one liter of fuel, and it was both handsome and well built. Mercedes realized that the running gear of a Diesel automobile should match the engine's reputation for longevity. They later proudly publicized the record of a Swabian mover who transported cattle in a trailer towed by his 260D. In thirteen years he logged 1.3 million kilometers, 250,000 of them with a single engine. That performance was one of many that helped to make the 260D the first practical and successful Diesel-powered automobile.

Daimler-Benz's experience with the 260D in the 1930s,

and with thousands of automobile Diesel engines built during World War II helped them to begin civilian Diesel production again in 1948. The most pressing need was agricultural machinery, and while Diesel oil was still relatively cheap, European farmers could not afford to run a 2.6-liter engine. Mercedes' response was to develop, in parallel, a 1.7-liter Diesel, and a gasoline engine of identical displacement. The two engines shared the same cylinder block, crankshaft, and many other components, and could be assembled on the same production line. The first 1.7Ds went into the versatile Unimog, which could be used as a truck, tractor, or general utility vehicle. In 1949, with a boost to 28 kilowatts at 3,200 r.p.m., the same engine was transplanted into the 170D sedan. The 170D sold for 1,200 Deutsche Marks more than its gasoline-engined counterpart, but with a top speed of 100 km/hr. and a fuel consumption of only 6.5 liters per 100 kilometers, it soon became an even greater success than the 260D.

Because of its historical associations, success for a Diesel engine has always included efficiency and longevity. One present-day counterpart of the 260D owner of the 1930s is a Pennsylvania machinist who has driven his 1953 170D sedan 1,270,000 kilometers—the equivalent of more than 31 times around the world. He drove the first million kilometers on the original engine, which was replaced when its oil pressure became a little low. The car was bought used in 1958 for $735 and it is still in daily use, averaging 15.52 kilometers per liter of fuel.

Until 1959 the history of the Diesel passenger car is essentially the history of Mercedes-Benz. By that year the 1.7 had been replaced with a new 1.9-liter engine, which developed 36.8 kilowatts—the same output as the original Benz truck Diesel of 1923, but from less than one-quarter of the displacement. Over the next decade this engine evolved into the 2.4-liter powerplant of the 240D sedan.

Meanwhile other manufacturers began to build Diesel-powered cars. In France, Peugeot's Lille engine factory had been in near-desperate straits since the end of World War II. The division found its salvation with the license production of a four-cylinder automotive Diesel for the Peugeot 403 sedan. The first engine came off the line in July, 1958, and by mid-1959 the factory, retitled with the acronym Indenor, was turning out 200 per day. Peugeot found the European market more than ready to absorb some competition for Mercedes; during the next eight years they sold 75,658 Diesel 403s.

The 403D's success prompted Peugeot to offer a Diesel option in the larger 404 sedan of 1962, and also sparked the interest of another French car manufacturer. In 1964 Citroën S.A. bought a joint interest in the Indenor engine plant, although the factory continued to make automotive engines solely for Peugeot.

Peugeot's automotive Diesels use a different combustion chamber from the descendants of L'Orange's vorkammer. It is called a swirl chamber, and one can almost feel how it works from its German name of *wirbelkammer*. The pioneer of swirl chamber design was Sir Harry Ricardo, a distinguished British internal combustion engineer who founded his own consulting company in 1919. Many of the swirl chamber designs used by Diesel engine builders like Peugeot are licensed from Ricardo and Company, Ltd.; the most popular one, perfected in 1933, is the Ricardo Comet Mark V.

The first obvious difference between a swirl chamber and a precombustion chamber is assymmetry. Precombustion chambers are usually placed above the center of the cylinder, and the fuel injector nozzle is mounted on the same centerline at the top of the combustion chamber. The typical vorkammer is round in cross section, and the ports from it to the cylinder are either on the central axis or spaced equally around it. The whole arrangement looks as though it could have been turned on a lathe.

By contrast the swirl chamber is always set off center. A section through the cylinder head of a swirl chamber engine reveals a compartment shaped like a three-dimensional number 9—a small hollow sphere with a vent tunneled to the edge of the cylinder space. The injection nozzle and the glow plug both open into the swirl chamber, and the fuel nozzle is usually at a shallow angle to the chamber wall on the opposite side from the cylinder vent. If you visualize the fuel injector and the cylinder port as BB guns, both of them would shoot little balls round and round the wirbelkammer in the same direction, but from opposite sides. That is exactly what happens first to the air compressed by the piston, and then to the fuel spray from the injector; the swirl chamber is a good mixer.

The combustion characteristics of the swirl chamber and the precombustion chamber are similar; both are quieter and slightly less economical of fuel than direct cylinder injection. The swirl chamber promotes perhaps the cleanest burning of a wide range of fuels, but it is harder to cold start than the vorkammer. The performance of each combustion system is very dependent on small changes in shape—there is still a lot of art in it—so many different designs have been proposed and tested. In 1972 a new swirl chamber automotive Diesel was put on the market to compete with Peugeot and Mercedes: the Opel 2.1 liter.

Like its competitors, Adam Opel AG has a long history of successful car manufacture. The company has been owned by General Motors since 1929, and during the 1930s it was the largest producer of automobiles in Europe. Opel entered the Diesel car field with a bang: it added a small turbocharger to one of its new 2-liter engines, boosting the output to 70 kilowatts at 4,500 r.p.m., and installed the package in a streamlined Opel GT coupe. On June 1–3, 1972, this car broke two world and eighteen international speed and distance records, including a flying kilometer at 197.498 km/hr. The same Diesel (unsupercharged) was offered as an option in seven of Opel's medium-sized cars, and it was priced significantly below the competition. The year it appeared, Opel displaced Volkswagen as the best-selling make of automobile in Germany.

In 1974 the degree of cooperation between Peugeot and Citroën became considerably greater, with Automobiles Peugeot S.A. acquiring 38 percent of Citroën's stock. The same year Citroën followed Peugeot's lead and brought out the fourth marque of Diesel-powered car. Like the Peugeots, the Citroën CX 2200D engine used a Ricardo Comet Mark V swirl chamber, but, surprisingly, it didn't share a single basic dimension with the Peugeot Diesel of almost identical displacement. In 1976, when the 2200D had been on the market for two years, Citroën was absorbed into the Peugeot group. In April of the same year Peugeot's Lille factory built its millionth Diesel engine.

Twenty

Perhaps it will help to clarify the current ferment in the automotive Diesel field by comparing three different designs. The first is a product of forty years' evolution, and one of the flagships of Daimler-Benz's Diesel line: the Mercedes 300D.

By 1974 the classic Mercedes passenger Diesel had been enlarged to a displacement of 2.4 liters, which Daimler-Benz now considered a practical upper limit for a four-cylinder automotive engine. In the past the next step for increased power has usually been to a six-cylinder design, in either in-line or V configuration. Instead, Mercedes chose to introduce in 1975 the unorthodox arrangement of five cylinders in line, each of the same volume as its predecessor in the 240D. The resulting engine has a displacement of 3005 cc, and develops 59 kilowatts at 4,000 r.p.m. While there have been five-cylinder truck engines, they are direct-injection designs in which a certain amount of noise and vibration are accepted. The six-bearing crankshaft of the 300D poses some interesting problems in forging and counterbalancing. In exchange for solving them, Daimler-Benz gained a short, relatively stiff engine that fits comfortably in the same bay as its smaller antecedents.

For anyone familiar with Mercedes products there are few surprises in the 300D engine. It uses the same high-grade materials and scrupulous details: the cylinder head is cast iron, the chain-driven camshaft is overhead, the valves rotate at every lift to equalize wear. The Bosch injection pump is the successor to a long line of proven designs, and it distributes fuel to a set of precombustion chambers that Prosper L'Orange would recognize at a glance. The engine starts and stops with the turn of a key (a convenience that takes some engineering in a Diesel) and it runs with impressive flexibility and smoothness. It is offered in a deluxe version of Mercedes' smaller sedan body, which it can propel at a maximum speed of 148.8 km/hr., and in a coupe and station wagon as well. Since its introduction there has been approximately a six-month wait for one in both America and Germany.

Nearly every second car that Mercedes builds is one of their four Diesel models, and the make as a whole commands a lion's share of the Diesel car market worldwide. But, in Mercedes' home territory in 1976, there appeared an audacious Diesel mouse (or in America, a Rabbit): the Volkswagen Golf. Like other competitors of Daimler-Benz, VW began by adopting a Richardo swirl combustion chamber. They next took a leaf from Mercedes' own book by designing a Diesel that could be manufactured on the same machine tools and assembly line as a twin gasoline engine. Daimler-Benz had done just that with the 170 series in 1948, but VW carried the principle of cross-compatibility even further. Of course there were advances in materials and technology in the thirty years between the two companies' projects; what is refreshing is that Volkswagen took advantage of so many of them.

The Golf Diesel is a four-cylinder in-line engine with a displacement of 1,471 cubic centimeters. Its narrow block is slightly thicker-walled than the gasoline version because of the increased pressure at the cylinder head. Although a Diesel develops higher forces in its connecting rods and higher bearing stresses in the crankshaft, both Golf engines use the same versions of these components. Even more remarkably, the aluminum cylinder head of the Diesel, where one would expect to find the most differences, has the same basic measurements as the gasoline one. The camshaft and valves are also the same, and all accessories on both engines are driven by similar toothed rubber belts. The name "Bosch" appears on the Golf's fuel injection pump, but it is simpler than the Mercedes', and it has a manual adjustment for advancing the injection timing after a cold start.

What does all this technology accomplish? The output of the Golf Diesel is 37 kilowatts at 5,000 r.p.m. (The smog-equipped U.S. version in the Rabbit produces 35.8 kilowatts at the same speed.) It is mounted transversely at the front of the very successful Golf sedan, and drives the front wheels. The DIN mileage rating for this combination is 6.5 liters per 100 kilometers. VW's introductory advertising emphasized that the Golf was the quickest accelerating Diesel car from zero to 100 km/hr. in production, a claim that remained valid until late 1977. It also has the longest oil change interval—7,500 kilometers—of any automotive Diesel marketed thus far.

There is a wait of approximately the same time as the Mercedes 300D for a Diesel Golf/Rabbit in Germany or the U.S.; since its introduction the order rate at the Wolfsburg factory has been running at 1,000 per day.

The third member of this trio is the newest, and it realizes Clessie Cummins' dream of an American-built Diesel automobile, forty-eight years after Cummins' brave experiment of 1929. This is the Oldsmobile Diesel, introduced in the autumn of 1977.

If cars can be said to have national characteristics, Oldsmobile's entry in the Diesel market must be as American as apple pie. It is a pushrod V-8 with a displacement of 5.7 liters—nearly twice that of its largest European competitor—and it produces 89.5 kilowatts at 3,600 r.p.m. Like Volkswagen, Oldsmobile began with a gasoline engine of identical displacement to its eventual Diesel, and spent three years developing and testing the necessary changes.

The engine that Oldsmobile evolved is in the United States automotive tradition: big, beefy, relatively slow running, and favoring quietness and acceleration over economy. In addition to changing the pistons, Olds has enlarged the crankcase webs and bearings, and substituted heavier connecting rods for the ones used in the gasoline-fueled prototype. The Oldsmobile Diesel is a swirl-chamber engine; for the all-important cylinder head GM's engineers first tested a Ricardo combustion-chamber design, but rejected it as too noisy. The shape they evolved is an inverted dome near the outboard edge of each cylinder, and it does indeed give unusually quiet combustion. Fuel is injected through a rotary pump made by Roosamaster division of Stanadyne, Inc., in Hartford, Connecticut. The Olds combustion chambers are preheated by electric glow plugs, and the engine has proven to be relatively quick to cold-start.

Installed in one of the big Oldsmobile sedans, the new Diesel looks little different from a GM gasoline engine. It is surrounded by the auxiliaries that American drivers have come to expect, plus a special hydraulic booster for the brakes and a vacuum pump for accessories. The car drives like any other Oldsmobile, with similar throttle response, the highest acceleration of any Diesel passenger car yet marketed, and no trace in the soft ride of the extra 82 kilograms of weight up front. Oldsmobile claims EPA fuel consumption ratings of 11.76 liters per 100 kilometers in city traffic and 8.11 liters/100 km on the highway, but it will take a while before extended driving records confirm or alter these estimates.

Except for the mandatory oil change at 5,000-kilometer intervals, and looking for a different set of service stations, owning the Oldsmobile Diesel should be little different from owning its gasoline-fueled counterpart. Indeed, General Motors has made every effort to insure that that will be the case. In 1978 they are marketing the same Diesel engine in Chevrolet and GMC light trucks, and their engine sales target for the year is 100,000 units.

The size of that number and the diversity of the three passenger cars described above are indicators of a significant change. Historically the automobile has absorbed a very small portion of the total Diesel engine market. During the first 60 years of the engine's existence, only one company sold series production Diesel cars, and in 1976, when about 5,000,000 Diesel engines of all sizes were built in the Western world, a mere 6 percent of them went into automobiles. To give a more precise perspective, in 1976 Daimler-Benz built 370,000 passenger cars worldwide; 43 percent were Diesel-powered. During the same year Mercedes sold 43,205 automobiles in the United States, of which 46.4 percent had Diesel engines. But that percentage had risen from 14.4 to 46.4 between 1972 and 1976.

The same increase of interest is reflected in statistics from many other countries. In Sweden, Volvo has contracted to buy 20,000 Volkswagen Diesel engines for its 1978 240 series sedans. Fiat, Alfa-Romeo, and Renault are joint partners in the recently founded Society of French and Italian Motor Companies, which plans to build automotive Diesels with outputs ranging from 40 to 115 kilowatts for its member companies. Saab, Peugeot, and Ford of Europe have expressed interest in buying engines from SOFIM, and companies like Audi and BMW, which have never marketed Diesel-powered cars, are now developing Diesel models. Nissan, until recently the only domestic producer of Diesel automobiles in Japan, quadrupled its output of them between 1975 and 1976 and is continuing to build up production. Honda, Toyo Kogyo, and Toyota are all developing Diesel cars, and Mitsubishi has sold its four-cylinder Diesel engine to Chrysler for installation in Dodge pickup trucks.

Why the sudden interest? First, and most important, efficiency. In a way it is a relief to have Rudolf Diesel's dictum finally driven home to the man in the street; it is only unfortunate that it has taken a world energy crisis to do it. The conditions can be stated simply, but with enough truth to justify the simplification: Petroleum has been a primary fuel of the world industrial complex for the last century. The Diesel engine is the most efficient petroleum-fueled prime mover thus far invented. Users of industrial prime movers were the first to recognize this, and continue to be the most intensive users of Diesel engines. Private automobile users, especially those whose standard of living has been raised the most by industrialization, are the world's greatest wasters of petroleum. Now, suddenly, it has been forced upon their reluctant attention that petroleum is running out and it is important to use fuel more efficiently.

Independent tests made recently in England and Germany showed that at the same speed, with the same load, a Diesel-engined automobile uses 25 percent less fuel than a gasoline-engined one of the same size. The tests conducted by ERDA's Energy Research Center at Bartlesville, Oklahoma in early 1977, including the 1975 Federal Test Procedure and the EPA standard mileage tests, as well as cruising comparisons at various speeds, confirm this conclusion. Comparisons were made between identical American models powered with the same size gasoline and Diesel engines, and even by substituting a Diesel for a gasoline engine in the same chassis. The results showed that in medium-weight passenger cars a Diesel engine used 43 percent less fuel than the same size gasoline engine for the same test cycle, and in heavier vehicles the Diesel used 35 percent less fuel than the gasoline engines. It is not a trick or a qualification; it is inherent in Rudolf Diesel's original premise. The rising percentages of Diesel car sales are first of all an indication of long-overdue efficiency awareness among automobile buyers.

A second factor in increasing Diesel car production can also be traced back to Rudolf Diesel: air pollution. In the nineteenth and early twentieth centuries that meant smoke and noxious gases from coal-fired furnaces. Any visitor to London before the 1950s can confirm that the individual citizen was just as effective at pollution with coal as he is with gasoline. The prime targets of antipollution programs are unburned hydrocarbons, carbon monoxide, fuel vapor emissions, nitrous oxide, and sulfates. Because the Diesel cycle uses a surplus of air, its combustion is typically more complete than the gasoline engine's. As a result its unburned hydrocarbons and carbon monoxide emissions are far below those of gasoline engines without any emission controls at all. The same efficient combustion reduces sulfate emissions to less than a quarter of those put out by a catalyst-equipped gasoline engine. The vapor pressure of Diesel oil is near zero at normal temperatures, so that, unlike gasoline, it also requires no vapor emission controls.

In four of the five air-pollution classifications listed above, the Diesel is effortlessly superior to the gasoline engine. The fifth one, nitrous oxide, is more difficult to control in a Diesel because the same high temperatures that produce efficient combustion are also likely to produce oxides of nitrogen. Normally charged Diesels are, in the opinion of many engine designers, near the limit of efficient operation at the Environmental Protection Agency nitrous oxide specification of 0.93 grams per kilometer. Diesel engines also produce some emissions that, while not classified as pollutants by federal agencies, ought to be in the opinion of many drivers. These include a fallout of black particles, mostly pure carbon, that result from rich combustion, and an unpleasant odor from aldehyde gas components in the exhaust. Daimler-Benz and Volkswagen have both recently stated that these unclassified emissions can be greatly reduced or even eliminated. The same two companies also indicated late in 1977 that a more dramatic present for the car buyer was just around the corner: an even cleaner-burning, more economical, and much more powerful automotive Diesel.

Twenty-one

On June 15, 1976, in the midst of a surge of automotive Diesel competition, and perhaps some journalistic rumbling that its products were falling behind the market, Daimler-Benz unveiled a surprise: the fastest Diesel car in the world. In 64 hours on the Nardo racecourse in southern Italy, a Mercedes C-111 broke three world and thirteen class speed and distance records previously held by gasoline-powered cars. To accomplish this the car averaged 253.030 kilometers per hour for 24 hours, and covered 16,000 kilometers at an average speed of 251.798 kph.

Certainly the sleek C-111 has looked the part of a record-breaker ever since Mercedes demonstrated it with various versions of the Wankel rotary engine. When the rear deck was lifted after this record run, the engine bay looked surprisingly empty and almost boringly conventional. The powerplant was a 300D Diesel engine, looking as though it had been plucked out of a production line sedan—except that it was equipped with a small Garrett Airesearch turbocharger. The combination produced more than 147 kilowatts at 4,200 r.p.m. from only three liters of displacement.

Volkswagen, not to be outdone, soon released figures on its own turbo-Diesels: Two VW Golf sedans fitted with turbochargers could accelerate from 0–100 kilometers per hour in less than 15 seconds, and their combined city/highway fuel consumption averages 4.7 liters per 100 kilometers. Both Volkswagens surpassed the strict EPA pollution standards projected for 1981 in everything but nitrous oxide. These are the kind of claims that provoke immediate skepticism in many knowledgeable automobile buyers, but in this case the claims are correct. Daimler-Benz began to market its 300SD supercharged Diesel passenger cars in the spring of 1978, with VW not far behind. The commercial proving time for these cars should be short because the technology is neither radical nor new; supercharging is as natural to the Diesel engine as breathing.

Rudolf Diesel thought of air not only as a diluent for fuel and a source of oxygen for combustion, but as a thermodynamic working fluid. As a result air compression has been linked with the Diesel engine from its beginnings. The first Diesel, a four-cycle, required auxiliary compressed air for starting and fuel injection. Its piston, designed for high compression, was just as effective at lowering the cylinder pressure during the intake stroke, thus allowing air at atmospheric pressure to flood into the cylinder for the next cycle. This continued to be the charging method for the four-cycle engines that followed, but the situation changed with the development of the two-stroke Diesel. A two-cycle engine doesn't have the luxury of a separate intake stroke, and its intake and exhaust ports are open simultaneously while the cylinder is cleared of burnt gases and fresh air is brought in. Normal piston aspiration is unable to supply enough air under these conditions. Several forced charging methods were invented or adapted from the two-stroke gasoline engine to satisfy the breathing requirements of the Diesel. They included separate air pumps, and crankcase scavenging, in which the underside of the main piston pressurizes the air in the crankcase on the downstroke and forces it up into the cylinder through transfer passages.

Improvements in these air supply systems moved in step with improvements to the engines, and it was soon realized that it was impossible to make the air pump too efficient. In a Diesel, the more fresh air the better; it means a denser charge, hence higher output; cleaner burning, because exhaust gas is expelled more completely; and longer-lasting components because the engine runs cooler. During the early decades of the twentieth century reciprocating air pumps were built into most two-cycle Diesels. As engines became larger, more air was needed, and engine-driven centrifugal blowers began to be used to charge some Diesel two-strokes. The conceptual jump from charging with a surplus of air at atmospheric pressure to charging with precompressed air at higher than atmospheric pressure—supercharging—had been made already; a supercharged Diesel engine was running at Gebrüder Sulzer in 1912.

That project was started by a man whose name is still a byword in supercharging: Alfred Büchi. He began his experiments on a four-stroke engine, but the high-pressure goals he aspired to proved too far ahead of their time. Büchi continued to work on the problem, and by 1924 several companies were building supercharged two-cycle Diesels based on his research. It will come as no surprise by this time to learn that supercharging followed a leapfrog alternation from two-stroke to four-stroke and back again. By the 1950s many manufacturers were offering stationary and marine Diesels in two- and four-stroke models with supercharging.

Superchargers are simply air pumps, and they fall into two main groups: Pumps driven by the engine or an outside motor, and pumps driven by exhaust gases impinging on a turbine, or turbochargers. Mechanically driven superchargers can include reciprocating pistons, several kinds of positive displacement blowers, and centrifugal compressors. Exhaust turbochargers usually have centrifugal compressors. Each of these types has definite advantages and drawbacks. The most common positive displacement blowers, for example, use a pair of rotors that look like rounded-tooth gears meshing inside an airtight housing. Inlet air is trapped between the "teeth" of the rotors and forced into a smaller volume as the lobes rotate. Such pumps (often called Roots blowers after their original producer, the Roots-Connersville Corporation of Connersville, Indiana) deliver nearly the same amount of air per revolution regardless of engine speed, and they are used in applications where speed changes are frequent. They are limited to a pressure ratio of about 1.5:1. They also require a mechanical drive, generate a characteristic throbbing noise, and can absorb considerable power.

Exhaust-driven turbochargers have no mechanical connection to the engine except a gas duct. They harness power that would seem to be otherwise wasted by using the hot exhaust gases to spin a turbine rotor. The rotor is on a common shaft with the centrifugal blower that supplies compressed inlet air to the engine. Turbocharged engines can operate at higher pressure ratios with lower specific fuel consumption than unblown engines, but as in all real-life engineering, this improvement is not free. The energy necessary to spin the turbine is reflected in increased back pressure that the engine must work against. Turbochargers operate at high temperatures and very high speeds, typically from 20,000 to 100,000 r.p.m., so they must be strong and precise. Their air output is not constant, but varies as the square of their speed. That means that at low engine speeds they supply relatively small amounts of air. Since they have no direct connection to the engine, their response to sudden changes in speed or load tends to lag the engine's needs. They too generate considerable noise, at higher frequencies than positive displacement blowers.

Most of the drawbacks described above (and there are others) can be tempered by careful design and selection. Overall, the benefits of supercharging Diesels far outweigh the drawbacks. The proof is self-evident in the worldwide adoption of supercharging for both two-stroke and four-stroke designs during the past twenty years, and in the continuing pace of research on supercharging. Some outstanding recent experiments include the Perkins differential supercharging system that has a nearly constant power output over a wide speed range, and Caterpillar's test engine with the very high supercharger pressure ratio of 9.0:1.

Perhaps the most radical breakthrough in this area is the Comprex pressure wave supercharger developed by Brown, Boveri and Company of Switzerland. The concept of a gas pressure exchanger goes back to a German patent of 1913, and much of the research on the present machine has been done under the supervision of Professor Max Berchtold at the Swiss Federal Institute of Technology in Zurich. The Comprex PWS is an engine-driven supercharger that uses the energy of the exhaust gases not to drive a turbine, but to compress incoming air. It is a deceptively simple-looking machine with only one moving part, a hollow cylinder mounted on an axle and divided by wavy vanes connecting the central shaft to a thin outer wall. In crossection this rotor looks like a baroque spoked wheel, and you can see straight through it from one end to the other. It is mounted in a housing that admits air at atmospheric pressure and engine exhaust gas, and puts out compressed air to the cylinders and exhaust gas at low pressure. Obviously, what happens inside the simple rotor is very complicated; even a brief description of the gas dynamics involved fills quite a few pages of formulas and graphs. What is important is that it combines some of the best features of a positive blower and a turbocharger without their disadvantages, and that its mechanical drive uses only about 1 percent of the engine power. It has been tested with great success on vehicles ranging from passenger cars to heavy trailer trucks in Switzerland and Germany, and on a White truck powered by a Cummins Diesel in the United States. Brown, Boveri is now marketing six sizes of the new supercharger for engines of 100 to 315 kilowatts.

The Comprex PWS is an example of the novel research that the Diesel engine has provoked during its short life. There is a vigorous quality about it that challenges the ingenuity of designers and engineers, and it is interesting to compare some of the engines currently being built by the pioneer Diesel manufacturers with their ancestors.

During the 80 years since the tests of the first Diesel engine, M.A.N. has grown into a conglomerate with twelve divisions and annual sales of DM 5,000,000,000. In 1976 they exported machinery to fifty-nine countries, and their 1977

Diesel line included 140 different engine models in eight series. Forty-two of these designs are two-stroke, the remaining ninety-eight four-strokes. Rudolf Diesel and Heinrich Buz were elated when Professor Schröter's tests showed that their 1897 engine produced 13.1 kilowatts. The smallest M.A.N. Diesel today is a six-cylinder marine auxiliary engine built at Nürnberg. It is barely over 1 meter long and develops 51 kilowatts at 1,500 r.p.m. The other end of the size spectrum looks quite different. M.A.N.'s largest Diesel, a two-stroke turbocharged engine for marine and stationary applications, is built at the Augsburg plant. It has twelve cylinders in line, is 22.5 meters long and stands 10.8 meters high—about three stories. Its power rating is 32,400 kilowatts at 110 r.p.m., and it weighs 1,150 tonnes.

Gebrüder Sulzer, where Rudolf Diesel began his engine-building career, has a similar history. The Sulzer Group now includes ten divisions producing a wide variety of heavy and precision machinery. In 1976 sales totaled 3,529,000,000 Swiss Francs, and the company employs a total of 35,000 *blaue Monteurs* and other workers. Diesel engines remain a major part of Sulzer's business, with a range of sizes similar to M.A.N.'s. Sulzer's 144 Diesel types are divided in a way that reflects their early achievements in two-cycle design: 82 two-strokes and 64 four-strokes. The Swiss have not lost their fascination with size. The largest Sulzer Diesel is the world's best-selling low-speed turbocharged two-stroke. It produces 35,280 kilowatts at 108 r.p.m., and at 29.23 meters long and 13.34 meters high it is the size of a large chalet. Sulzer gives its dry weight as 1,686 tonnes.

Perhaps it has something to do with living in a small country. Burmeister & Wain, the Danish company with which Rudolf Diesel had such cordial rapport, has also prospered in the intervening years. At the top of their line of 156 Diesel models is a huge two-stroke engine like those made by M.A.N. and Sulzer, with a maximum rating of 35,900 kilowatts at 103 r.p.m. B & W lists this awesome prime mover as being designed specifically for marine propulsion. That field also offers an appropriate example of B & W's progress since it introduced the seagoing motorship in 1912.

In April of 1972 Burmeister & Wain launched a new *Selandia* for the same customer that bought the original one, the Danish East Asiatic Company. Like its famous predecessor, this ship is all Danish, built in B & W's dock and powered with B & W engines. She is two and a half times longer and twice as wide as her namesake, and her gross tonnage is five times larger. She is able to move her all-containerized cargo at 50 kilometers per hour, 2.5 times the speed of the original *Selandia*. While it is relatively straightforward to increase the size of a ship, it is both more difficult and more expensive to increase its speed as well. *Selandia's* engine room, sixty years later, is again unique.

The new ship is driven by three turbocharged two-stroke Diesels, a center engine with twelve cylinders, and two flanking engines with nine cylinders each. The main engine develops 22,080 kilowatts at 117 r.p.m., and the side engines produce 16,560 kilowatts each at the same speed. The total of 55,200 kilowatts is the largest marine Diesel installation in history. The utilization of this power is also new and unique: the center engine drives a five-bladed adjustable pitch propeller, and each wing engine is coupled to a six-bladed propeller that can be disengaged from its shaft. On long ocean passages all three engines are used; between local ports the outer propellers are allowed to rotate freely while the center engine alone drives the ship at a fuel-saving 40 kilometers per hour.

Clearly B & W feels an obligation to continue its tradition of marine innovation. Still, in one respect the new *Selandia* may be said to have retrogressed. Though she is equipped with radar, air conditioning, and servo-controlled stabilizers, and is even longer and sleeker than her ancestor, just aft of her bridge is a pair of tall, bright yellow streamlined funnels.

Twenty-two

Size is not the only measure of progress. Any machine developed by a responsible company or individual is tested and modified many times before it reaches the market, but from the time it enters service more changes become necessary. It is impossible for even the most conscientious designers to foresee every use and abuse of a product, or to anticipate revisions based on practical experience. These factors insure that a new design is seldom an ultimate design—there is always room for improvement.

The Diesel engine provides a superb example of design evolution and improvement. Its birth was difficult, and its initial market acceptance was slow, so manufacturers felt impelled to put their best efforts into the development of the engine once they had made a commitment to it. The most important measure of the Diesel's success to its inventor was efficiency. The 26 percent thermal efficiency of the 1897 Augsburg engine was Diesel's confirmation that his concept was sound. By the 1930s several manufacturers were building Diesels with efficiencies of 34 percent—double that of contemporary gasoline engines. Since then the Diesel has been refined to the point where some direct-injection engines can reach a thermal efficiency of 40.5 percent. That is far higher than any other mobile prime mover, and would doubtless make Rudolf Diesel smile.

Marine Diesel engines are generally larger, their weight is less critical, and they have an unlimited supply of sea water for cooling and the recovery of waste heat. The overall efficiency of Diesel marine plants making use of these advantages is close to 50 percent, and a few installations have passed that magic number. The reciprocating steam engine has been left far behind as a competitor in this area, and even the largest marine steam turbines with high pressure reheat can reach only about 36 percent efficiency at the present time.

In all types of Diesels, the steady increase in efficiency has been accompanied by lower specific weights and better fuel economy. The amount of fuel burned per unit of power produced is the engine user's practical measure of efficiency. It is probably the most important reason for the Diesel's long-term success, and for the new surge of interest in it. In a sense the engine has an eighty-year head start over other prime movers because of Rudolf Diesel's conviction that low fuel consumption was an essential virtue. Diesel engine manufacturers were selling fuel economy in the days when it was almost ignored by users who felt that oil was inexhaustible. Now they have a machine that produces more power per unit of fuel than any other, and we know all too well that oil is not inexhaustible. But, however efficient the Diesel is, it is just as dependent on its fuel as any other prime mover. What about Diesel fuel?

To begin with it isn't fuel, but fuels, just as there are different types and grades of gasoline. Diesel fuels are oils distilled from crude petroleum; they are heavier than gasolines, and close to kerosene in distillation point. The Federal Government and the American Society for Testing Materials (ASTM) grade four Diesel oils, the middle two of which make up the bulk of the Diesel fuel sold in the United States. (The others are an arctic winter fuel, and a heavy distillate burned in large stationary and marine engines.) Twelve properties of these fuels are specified in the ASTM classification. They include ash content, carbon residue, cetane number, corrosion index, flash point, heating value, pour and cloud points, specific gravity, sulfur content, viscosity, volatility, and water and sediment content (or cleanliness). All of these are important to some degree, but the ones most important to the consumer are cleanliness, low temperature fluidity, and cetane number.

Cleanliness is critical in Diesel fuel. Fuel pumps and injectors are made to extremely close tolerances, and they can be irreparably damaged by any form of dirt or excess water. Most American refineries, to their credit, have a near-fetish about keeping contaminants out of high grade Diesel oils. It is obvious that if fuel won't flow, the engine won't run, hence the importance of low temperature fluidity. The simplest measure of fluidity is the pour point, and the related cloud point, usually about 5.6°C higher. The pour point is simply the temperature at which the fuel stops flowing; the cloud point is the slightly higher temperature at which wax crystals begin to condense out of the fuel, blocking the filters and stopping the engine.

Cetane number is a measure of a Diesel fuel's ease of ignition. Normal cetane is a hydrocarbon with the formula $C_{16}H_{34}$; it ignites very easily with a short delay time. The opposite characteristic of slow, poor ignition is shown by another hydrocarbon, alpha-methylnaphthalene (α–$C_{10}H_7CH_3$). A reference fuel of 100 percent cetane was assigned the cetane number 100; 50 percent cetane and 50 percent alpha-methylnaphthalene the cetane number 50, and so on. When Diesel fuel is tested, it is assigned a number equal to the percentage of cetane in the reference fuel with the same ignition characteristics. Most modern high-speed Diesel engines require fuels with cetane numbers higher than 40, and will barely run on fuels of 25 cetane and less.

Diesel Grade 1-D is a premium quality, relatively volatile fuel, usually distilled from selected crude stocks. It is allowed only minute traces of water and sediment, a minimum cetane number of 45, and a pour point that can be adjusted for temperature range, but that is significantly lower than that of other Diesel oils. It is typically light in color, and it is the fuel of choice for high-speed engines that are sensitive to cetane number, or subject to frequent variations in speed and load.

Grade 2-D is a heavier, usually amber-colored fuel with a higher viscosity and pour point than 1-D. It has a minimum cetane number of 40, and is allowed slightly higher amounts of impurities than 1-D. It is considered a general purpose fuel for all Diesels up to the very large engines which will burn Grade 4-D. All of these grades have heating values—energy content—at least 10 percent higher than gasoline. The ASTM cetane standards are exceeded as a matter of course by American refineries, whose Diesel fuels usually have cetane numbers in the high 40s.

Beginning around 1961 detergents and dispersants began to be added to Diesel oils to lessen deposits and extend the life of injector parts. These were followed by cetane improvers, which are effective but costly. Other additives can include rust, corrosion, and gum inhibitors, pour point depressants, and antismoke compounds. (These chemicals are seldom all present in a single fuel.)

How much Diesel fuel is made? Standard Oil Company of California manufactures some 36 million kiloliters of salable fuel oils per day, and this represents approximately 12 percent of their daily crude oil run. Of the fuel oil total, 1.7 million kiloliters is Diesel 1-D, and 25.1 kiloliters is Diesel 2-D. About half of all the Diesel fuel is sold to heavy industrial and agricultural customers, 30 percent to road transportation accounts, and the remaining 20 percent to the railroads. Most of these customers buy their fuel in bulk, and the prices they pay for it have risen about 83 percent since 1973.

American oil companies differ sharply in their internal estimates of the cost of producing Diesel fuels relative to the cost of making gasolines. Mobil has made a study that concluded it could save nearly 50 percent of the gasoline refinery energy cost per unit of Diesel oil. By contrast Amoco's research and development department issued a paper in September of 1977 that was extremely conservative. Amoco refineries are currently producing about 0.7 liters of Diesel fuel for each liter of gasoline. Their prediction is that from this 0.7 to 1 ratio up to a ratio of 1 to 0.9 (more Diesel fuel than gasoline), there is a potential energy saving of perhaps 1 percent, and a loss above that ratio. Standard Oil of California's appraisal lies somewhere between the two above, but closer to Amoco's position. These estimates are of course subjective, and may reflect differences in refinery efficiency and company policies, as well as the attitude of a conservative industry. At this point, even at retail, the combination of greater efficiency and a moderately favorable fuel price relative to gasoline makes the Diesel engine an unbeatable transportation value.

There are about 180,000 service stations in the United States, and an estimated 18,000 of them sell Diesel fuel. The price differential nationwide ranges from 1.0 to 2.5 cents per liter less than gasoline, and state taxes average about 1.25 cents per liter less; the federal tax of 1.05 cents per liter is uniform nationwide on both gasoline and Diesel oil. This narrow price spread makes an interesting contrast with the situation in other parts of the world.

Between April 9 and May 3, 1975, two members of the British 1975 Everest expedition and three professional drivers drove from Leeds, England, to Kathmandu, Nepal, a distance of 11,259 kilometers. After long deliberation, Chris Bonington, the leader of the expedition, decided that this was the best method of getting the required supplies to the Himalayas on schedule. The vehicles were two British Ford 16-tonne lorries, each powered by a 6.2 liter Diesel engine; both trucks were loaded approximately 2 tonnes overweight. Over the entire distance the trucks averaged 4.03 kilometers per liter of fuel, and their average fuel cost was $.0317 per kilometer. This is an excellent performance (and they arrived on time and succeeded on the southwest face of Everest as well), but what is also interesting is the range of prices paid for Diesel fuel during the trip. The highest price was in Germany: $.314 per liter. The same grade of fuel in Iran cost $.027 per liter. That is a retail multiple of 11.6 for the same commodity—free market pricing with a vengeance. The United States, the world's largest consumer of petroleum fuels, now has to face the unpleasant prospect of a future in which it has little control of that commodity market.

Twenty-three

Despite its tremendous consumption of oil, the United States has historically lagged far behind the rest of the industrialized world in the use and production of Diesel engines. That is still true today. Some 520,000 Diesels were built in America in 1976, about 10 percent of the free world production of 5,000,000 engines. (The Russian bloc countries and China are also heavily Dieselized, but it is difficult to get accurate use and production figures for those areas.) All United States manufacturers combined add about 105,000 new Diesel-powered trucks and buses to our vehicle fleet annually; overseas Daimler-Benz alone builds more than 140,000 per year, and that represents less than 15 percent of the total output for Europe and Japan. In the industrial and marine field, 82 percent of the engines built in Western Europe are Diesel, compared to 7 percent in the United States.

The same relationship is apparent in variety and size. Major American producers of Diesels include companies like Allis-Chalmers, Caterpillar, Cummins, John Deere, Detroit Diesel Allison, Fairbanks Morse, General Electric, International Harvester, Mack, and Perkins. Fairbanks Morse, now a division of Colt Industries, was one of the first American Diesel manufacturers. Their line today includes ten opposed-piston two-strokes with outputs of 450 to 2,950 kilowatts, and four four-strokes built under license from S.E.M.T.-Pielstick of France, ranging from 4,300 to 8,730 kw. Caterpillar, with some of the most modern Diesel production facilities in the world, markets twenty-three basic engines in various configurations with powers of 64 to 1,157 kilowatts.

The largest Diesel engine currently made in America is the Delaval RV-20-4 built in Oakland, California. It is a four-stroke, turbocharged V-20 that develops 10,100 kilowatts at 450 r.p.m. It is 8.35 meters long, stands nearly 5 meters high, and weighs 128 tonnes. It would be considered a medium-power, medium-sized engine in Europe, where a number of companies make directly competitive lines of V-type engines. The Enterprise Division of Delaval, which produces the RV-20-4, built its first Diesel engine in 1920, and its output of 350,000 kilowatts in 1977 places it among the smaller American Diesel companies. Like its co-manufacturers, it is used to selling in a market that believed it had an unlimited supply of gasoline. There are many signs that this situation is changing rapidly.

In 1968 the City of Los Angeles Fire Department, with some soul-searching, bought its first Diesel-powered truck. During the next few years it acted out the London Transport story all over again. As of mid-1977 it had purchased eighty-eight pieces of Diesel-engined fire equipment, and had thirty more on order. Having tested Diesels for nine years, the department now buys nothing else.

Many American cities have made similar reappraisals, but it is ironic to report its happening in Milwaukee, Wisconsin. The products that turned the tide in this cradle of American Diesel manufacture were Ford heavy-duty trucks powered by Caterpillar engines. *Diesel & Gas Turbine Progress*, one of the bibles of the industry, is published in Milwaukee. Its forecasts for Diesel sales and production in the United States are as rosy as the *Maschinenfabrik Augsburg* stock futures of 1897, and considerably more reliable. Its predictions are based in part on the fact that more than $1 billion has been invested in added production capacity during the past several years by American Diesel manufacturers.

One of Rudolf Diesel's most cherished ideals was the free flow of concepts and inventions across national boundaries. The idea of international cooperation was inextricably bound up with the Diesel engine in his speeches and writings. It is likely that the coming of age of the Diesel in America will be accompanied by an increase of internationalization. A number of United States Diesel manufacturers have had overseas factories for years; some recent examples of the reciprocal international trend include AMC's decision to install a perkins Diesel engine built in Hannover, West Germany, in its CJ Jeeps, John Deere's establishment of a manufacturing plant in France, and Hawker Siddeley's purchase of 36 percent of Onan Diesel's stock from Studebaker-Worthington.

The exposure of engineers to designs different from their own is often a beneficial experience. In 1977 Murphy Diesel of Milwaukee agreed to market engines built by Motoren-Werke Mannheim AG of West Germany. (MWM is the original stationary engine division of Benz & Cie that split off from the parent company in 1922.) Another international agreement may give the United States something it has never had before: Manufacturing experience with really large internal combustion engines. The marine department of Westinghouse Corporation in Sunnyvale, California, was recently licensed to build Sulzer's giant two-stroke Diesels of up to 29,580 kilowatts. The interchange of language and expertise that is inevitable in these transactions may help to realize some of Rudolf Diesel's nonengineering ambitions.

Given such encouraging prospects, it is tempting to view the Diesel engine as a panacea. Unfortunately it is no more a cure-all for human or environmental problems than any other device. On the level of simple mechanics it has all the failings of manmade artifacts. It is still too heavy, too noisy, too thirsty, too smelly, too hot, too dirty, and too expensive to be the perfect prime mover we would like to have. Even if it is made with exquisite care and workmanship, there will always be leaks and misalignments, worn bearings and cracked pistons. But it is just as subject to improvement as any other invention, and many of its imperfections can be lessened. It has a set of abilities that no other engine can match: It is tough; it is long-lived; it will tolerate an incredible range of fuels from kerosene to coal dust; it can be built light enough to fly airplanes and heavy enough to drive supertankers, and it has the most important advantage of all, an advantage that no country can afford to ignore: higher efficiency.

Some petroleum executives have recently issued statements that the Diesel engine will have to show a larger profit before they will adapt refineries to provide more fuel for it. They are wrong. As oil becomes scarcer, the real meaning of conservative will be forced on everyone. Saving must inevitably become as important as profit as supplies dwindle. Germany, with the highest price for fuel in Europe, has the highest level of Dieselization, and until a more efficient engine is discovered that relationship will be duplicated elsewhere. Certainly the United States has more to gain in this process than most other countries. If the sales of Diesel engines and Diesel-powered vehicles continue to climb at the present rate, you might be startled to drive up behind a car one day and see a new word on its trunk: GASOLINE.

How a Diesel Engine Works

The following examples are simplified descriptions of Diesel combustion cycles. While they are correct as far as they go, it should be remembered that in real engines the temperatures and pressures may be different from those mentioned, that there may be more than one intake and exhaust valve for each cylinder, and that real valve timing is more complex than an open-shut alternation, with extended periods of overlap. In single-cylinder engines like these models, the parts are kept moving during the nonpowered strokes by energy stored in the flywheel and crankshaft during the power stroke. In multiple-cylinder engines, the cranks are arranged to distribute the power impulses from the various pistons as smoothly and uniformly as possible through the cycle of rotation.

Four-Stroke Cycle

Historically, this was the first Diesel cycle. A complete four-stroke cycle requires two revolutions of the crankshaft, and the piston acts as a positive displacement air pump for two of its strokes.

1. INTAKE OR INDUCTION:

The cycle begins as the piston approaches the top of its stroke—top dead center—and the intake valve opens. The piston moves down, drawing fresh air at atmospheric pressure into the cylinder.

2. COMPRESSION:

As the piston passes bottom dead center the intake valve closes. The cylinder is now tightly sealed, and the piston

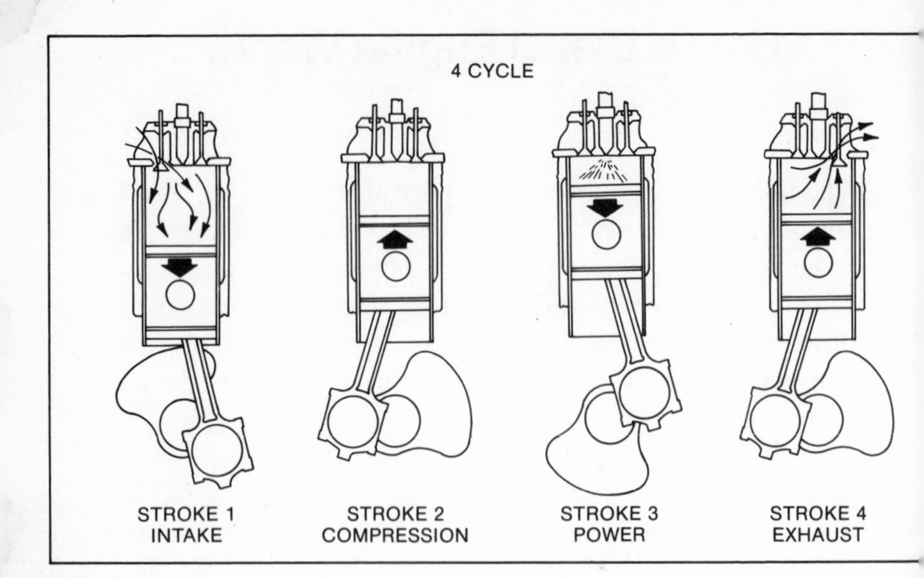

moves upward, compressing the trapped air to about 70 kilograms per square centimeter. The air temperature rises to some 500 °C by the end of this compression. When the piston nears top dead center again, fuel is injected into the combustion chamber.

3. POWER:

The fuel ignites, and the burning gases expand rapidly and with great force, driving the piston to the bottom of its stroke. As the piston moves downward, enlarging the combustion space, the pressure in the cylinder falls.

4. EXHAUST:

When the piston reaches bottom dead center for the second time the exhaust valve opens. The piston begins its upstroke, forcing the burned gases out of the cylinder, and as it nears the top of the stroke the exhaust valve closes, and the intake valve opens to begin another cycle.

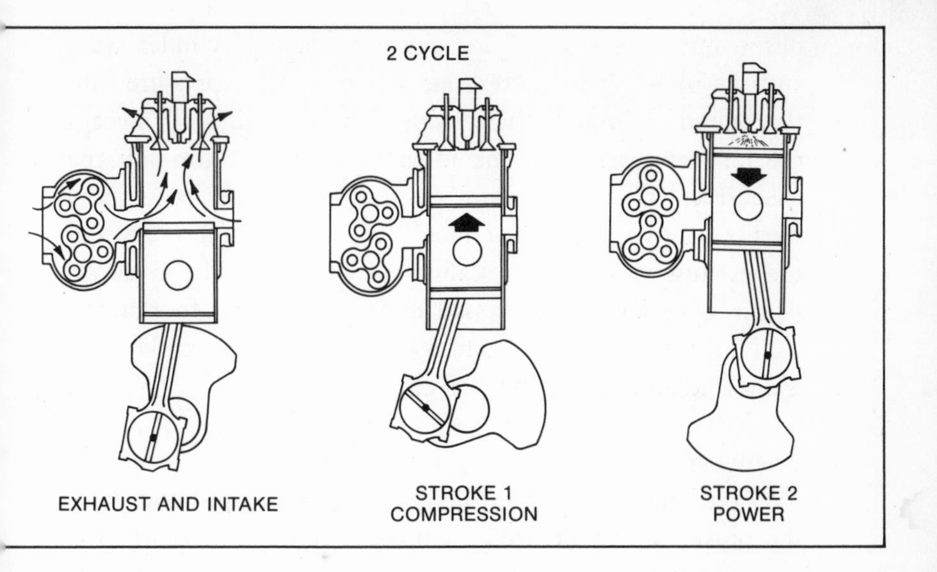

Two-Stroke Cycle

The two-stroke system was adapted to the Diesel engine in 1905, eight years after the first successful four-stroke Diesel. There are a number of variants of the two-stroke principle, including opposed-piston engines and so-called "uniflow" engines in which both the intake and exhaust gases flow in the same direction. A two-stroke cycle is completed in one revolution of the crankshaft, but when the piston is not used as an air pump, a separate blower is required to scavenge the exhaust gas and provide fresh air for combustion.

1. COMPRESSION:

The cycle begins with the exhaust valve open and the piston approaching bottom dead center. At this point the piston uncovers a set of transfer ports in the cylinder wall, and the blower forces fresh air at atmospheric pressure into the cylinder through these ports. The incoming air sweeps the burned gases from the previous cycle out through the exhaust valve. As the piston passes bottom dead center and starts to move upwards it blocks off the transfer ports and the exhaust valve closes, sealing the cylinder. The fresh air in the cylinder is compressed exactly as in the four-stroke engine, and when the piston nears the top of its stroke, fuel is injected into the combustion chamber.

2. POWER:

The fuel ignites, and the heated gases expand and drive the piston down. Before it reaches bottom dead center the transfer ports are uncovered, and the exchange of fresh air for exhaust gases begins again in preparation for another cycle.

Conversion Factors

The factors and equivalents below are meant as abbreviated aids for readers used to U.S. units of measurement. The rough conversions are just that; the tables are designed to allow more exact conversion, and to convey a sense of magnitude. For more complete data, *Instant Metric Conversion Tables*, issued by The Hamlyn Publishing Group, is a handy and easy-to-use set of metric-U.S. equivalents for many weights and measures. A comprehensive list of metric conversion factors for everyday, engineering, and scientific use is included in the current edition of *Machinery's Handbook*, published by Industrial Press, Inc.

1. LENGTH

Rough Conversions: There are approximately

25 millimeters in an inch	300 millimeters in a foot
2½ centimeters in an inch	30 centimeters in a foot
3¼ feet in a meter	⅝ of a kilometer in a mile

Millimeters	= Inches
1.0	0.039
10.0	0.39
25.4	1.0
100.0	3.93
254.0	10.0

Centimeters	= Inches
1.0	0.39
2.54	1.0
10.0	3.93
25.4	10.0
100.0	39.37

Meters	= Feet
0.304	1.0
1.0	3.28
3.04	10.0
10.0	32.8
30.48	100.0
100.0	328.0

Kilometers	= Miles
1.0	0.62
1.60	1.0
10.0	6.21
16.09	10.0
100.0	62.13
160.93	100.0

2. MASS

Rough Conversions: There are approximately
2 pounds in a kilogram. ½ kilogram in a pound.

Kilograms	= Pounds
0.45	1.0
1.0	2.20
4.53	10.0
10.0	22.04
45.35	100.0
100.0	220.46

3. VOLUME

Rough Conversions: There are approximately 16 cubic centimeters in a cubic inch. Four liters in 1 U.S. gallon (A liter is slightly larger than a quart).

Cubic Centimeters	=	Cubic Inches	Liters	=	Gallons
1.0		0.06	1.0		0.26
10.0		0.61	3.78		1.0
16.38		1.0	10.0		2.64
100.0		6.10	37.85		10.0
163.87		10.0	100.0		26.41
1000.0 (= 1 liter)		61.02	378.54		100.0

4. TEMPERATURE

Useful Equivalents:
0° Celsius is 32° Fahrenheit (Freezing)
20° Celsius is 68° Fahrenheit (Comfortable Room Temperature)
37° Celsius is 98.6° Fahrenheit (Normal Body Temperature)
100° Celsius is 212° Fahrenheit (Boiling)

°Celsius	=	°Fahrenheit
−17.8		0.
0.		32.0
10.0		50.0
20.0		68.0
37.8		100.0
100.0		212.0

5. PRESSURE

Rough Conversions:
1 kilogram per square centimeter is approximately 14 pounds per square inch
1 kilogram per square centimeter is nearly 1 atmosphere

Kilograms/Centimeter^2	=	Pounds/Inch^2
0.07		1.0
0.70		10.0
1.0		14.22
7.03		100.0
10.0		142.23
70.30		1000.0

6. POWER

Rough Conversions:
1 kilowatt equals 1⅓ U.S. horsepower
1 U.S. horsepower is approximately ¾ of a kilowatt

Kilowatts	=	U.S. Horsepower
0.745		1.0
1.0		1.34
7.45		10.0
10.0		13.41
74.56		100.0
100.0		134.10

7. SPEED

Rough Conversions:
25 miles per hour is approximately 40 kilometers per hour
55 miles per hour is approximately 90 kilometers per hour
8 knots is approximately 15 kilometers per hour

Kilometers/Hour	=	Miles/Hour
25.0		15.53
40.23		25.0
50.0		31.06
80.46		50.0
100.0		62.13
160.93		100.0

Kilometers/Hour	=	Knots (Nautical Miles/Hour)
10.0		5.39
18.53		10.0
20.0		10.79
30.0		16.18
37.06		20.0
55.59		30.0

8. FUEL CONSUMPTION

Rough Conversions:

5 kilometers per liter is about 12 miles per gallon

8½ kilometers per liter is about 20 miles per gallon

Kilometers/Liter	=	Miles/Gallon
4.25		10.0
5.0		11.76
8.50		20.0
10.0		23.52
12.75		30.0
20.0		47.04

Glossary

ATMOSPHERE—The pressure of the air at sea level. The standard unit—1 atmosphere—is the pressure that will support a column of mercury 760 millimeters high, at a temperature of 0° Celsius and a gravitational acceleration of 9.806 meters per second. One atmosphere is equal to about 1.03 kilograms per square centimeter, or 14.7 pounds per square inch.

ATMOSPHERIC CYCLE—An engine cycle in which the pressure of the atmosphere acts on some part of the engine to produce motion or useful work. This is usually accomplished by creating a partial vacuum—a region of lower than atmospheric pressure—inside the engine during part of the cycle.

BEAM—The width of a ship, measured across the hull at the point of greatest breadth.

BORE—The internal diameter of an engine cylinder.

CETANE NUMBER—A number between 0 and 100 which is a rating of the ignition quality of a Diesel fuel. The higher the cetane number, the easier the fuel is to ignite.

COMBUSTION CHAMBER—The space in which fuel is burned in an engine. It may be a rigidly bounded compartment, or it may be formed by moving surfaces which confine the burning temporarily. In a Diesel engine a precombustion chamber is one in which fuel is injected and partially burned before it reaches the main combustion space. A swirl chamber is a form of precombustion chamber which promotes rapid and turbulent mixing of compressed air and injected fuel.

COMPRESSION RATIO—The ratio of the maximum volume in an engine cylinder (when the piston is at its lowest point), to the minimum volume (when the piston is in its highest position). During the compression stroke the gas in the cylinder is compressed in proportion to this reduction in volume. The compression ratio in a typical gasoline-fueled automobile engine is about 8 to 1; in a similar-sized Diesel it is about 22:1.

CONNECTING ROD—The rigid beam, usually with a closed bearing at each end, that links the piston to the crank-shaft, and transfers the piston's motion.

CRANKSHAFT—The main rotary output shaft of a reciprocating engine. It has one or more offset sections—cranks—which receive the linear impulses of the pistons through connecting rods, and convert them into rotating motion.

CYCLE—In an engine, a sequence of actions which includes all the events necessary to sustain continuous operation, and which repeats itself identically. The most common sequences in reciprocating engines are the four-stroke cycle, and the two-stroke cycle. (See Appendix A for illustrated descriptions of these systems.) Engine cycles are sometimes named for their inventors, as Otto Cycle, or Diesel Cycle.

CYLINDER—The space in which the combustion cycle occurs in a reciprocating engine. It is ordinarily round in cross section as its name implies.

DISPLACEMENT—In a reciprocating engine, the total volume displaced by all the pistons in all the cylinders during their working stroke. If you were to set one piston at its lowest position and fill the cylinder with water to the level where the piston would be at the top of its stroke, the amount of water would equal the displacement of the cylinder. That volume of water for each cylinder in the engine poured together would equal the engine's displacement. (In fact that is how some competition engines are spot-tested for compliance with racing rules.)

The engine in a Volkswagen Beetle has a displacement of 1,584 cubic centimeters or 1.58 liters; a Cadillac Fleetwood engine has a displacement of 6,964 cc, or just under 7 liters.

DRAFT OR DRAUGHT—The vertical distance from the waterline of a ship to the lowest point of its keel; the depth of water necessary to float a ship, especially when loaded.

DYNAMOMETER—An instrument that measures, and often records, power output. Dynamometers are usually calibrated in watts or horsepower; they range in size from microbalances that can measure the output of a fly to water brakes which can absorb 50,000 kilowatts from a Diesel engine.

EFFICIENCY—The ratio of output to input. Brake efficiency is measured with a dynamometer that imposes an adjustable (and usually heavy) load on a machine by forcing it to work against a friction brake or a water brake. Mechanical efficiency is a measure of the frictional and mechanical losses in an engine. Thermal efficiency is the ratio of power output to the energy available per unit time from the fuel being burned.

INDICATOR—An instrument that graphs the pressure conditions inside an engine, while the engine is running. Preelectronic indicators look like miniature cylinder phonographs. The graph paper covering the cylinder is called an indicator card, and the power developed inside the engine—indicated power—can be calculated from the diagram traced on it.

INJECTOR—A device for injecting fuel into an internal combustion engine. Some injectors are complete self-contained pumps; others are actuated by remote pumps, or by camshafts or levers. All injectors must be able to meter small amounts of fuel precisely in a very short time, and must seal dependably against high pressures.

MAGNETO—An electric alternator, usually made with permanent magnets, that generates the ignition current for an internal combustion engine and is driven by the engine itself.

PISTON—The moving plunger that slides in the cylinder of a reciprocating engine. It must seal tightly against the cylinder walls, and still be able to slide back and forth with a minimum of friction.

POWER—The rate at which work is done. The SI unit of power is the watt. The earlier unit of power, the horsepower, was defined differently in metric and nonmetric countries. In the United States, 1 horsepower is defined as 550 foot-pounds per second, and equals 745.6999 watts. Conversely, 1,000 watts, or 1 kilowatt, equals 1.341022 U.S. horsepower.

PRESSURE RATIO—In a pump or supercharger, the ratio of output pressure to input pressure.

PRIME MOVER—A device that converts the energy in fuel into motion or useful work. A draft horse is a prime mover, and so is a rocket engine.

RECIPROCATING ENGINE—An engine in which the work-producing components (pistons, for example) move back and forth.

R.P.M.—Revolutions Per Minute. The number of times that a rotating shaft or machine turns in 60 seconds. For Diesel engines of different sizes, this number can be as low as 100, and as high as 5,000 or more.

SCAVENGING—The removal of burnt gases and other exhaust components from an engine combustion space.

SPECIFIC WEIGHT—In a prime mover, the weight of machine per unit of power produced, i.e., grams per watt, or kilograms per kilowatt. It is an indication of how efficiently the structural materials of the engine have been used by the designer.

STROKE—The distance traveled by the reciprocating parts of an engine during the power-producing segment of the cycle.

SUPERCHARGER—An air pump which is used to raise the pressure of engine intake air above atmospheric pressure. Piston pumps, rotary blowers, and centrifugal turbines have all been used as superchargers, and they may be mechanically driven by the engine they are supercharging, by exhaust gas turbines, or by separate motors.

TURBINE—A vaned wheel or assembly of wheels which spins as a result of gas or fluid impinging on it.

Books for Further Reading

Most of the biographical source material on Rudolf Diesel is in German, and very little of it is available in the United States. (In fact a surprising amount is in the archives of various German and Swiss companies.) There is one useful book-length biography of Diesel in print in America: *Rudolf Diesel, Pioneer of the Age of Power*, by W. Robert Nitske and Charles M. Wilson (University of Oklahoma Press, Norman, Oklahoma, 1965).

Much the same can be said of references on the early history of the Diesel engine, with the linguistic additions of French and Danish. There is, however, an excellent and comprehensive book in English on the origins of the internal combustion engine: *Internal Fire*, by C. Lyle Cummins, Jr. (Carnot Press, Lake Oswego, Oregon, 1976). This history starts with the earliest known internal combustion prototypes and ends with the first successful Diesel engine. Clessie L. Cummins, Patriarch of the same family of Diesel builders, published *My Days with the Diesel* in 1967 (Chilton Books, Philadelphia, Pennsylvania). Cummins' folksy reminiscence is too limited in perspective to be an accurate account of Diesel engine progress, but it is full of colorful vignettes of a struggling American industry in the 1920s and '30s.

Diesel and High Compression Gas Engines, by Edgar J. Kates and William E. Luck (3rd Edition, American Technical Society, Chicago, Illinois, 1974) is a practical technicians' introduction to the modern Diesel engine. This well-illustrated, American-oriented text uses United States units and engines throughout. A refreshing counterpoint to the cut-and-dried orthodoxies runs through *The Design of High Speed Diesel Engines*, by M. H. Howarth (American Elsevier Publishing Company, New York, 1966). This book manages to be both authoritative and personal at the same time, and it is particularly valuable because of its candor about the areas of engine design in which art and intuition still play an important part.

Definitive histories of the Diesel automobile, truck, and off-road vehicle still remain to be written in English, but the story of Diesel-powered ships up to the early 1950s is told in great detail in a *History of Motorshipping* by A. C. Hardy (Whitehall Technical Press, London, 1955). A cogent treatment of the Diesel and Diesel-electric locomotive with first-class illustrations is included in *The Concise Encyclopedia of World Railway Locomotives*, edited by P. Ransome-Wallis (Hawthorn Books, New York, 1959).

Index

C

D

E

F

G

H

I

J

K

L

M

N

O

P

R

S

T

U

V

W

Z